WORLD
CHECKLIST
AND BIBLIOGRAPHY OF

Magnoliaceae

WORLD
CHECKLIST
AND BIBLIOGRAPHY OF

Magnoliaceae

David G. Frodin

and

Rafaël Govaerts

First published October 1996

Address of the authors:
Herbarium, Royal Botanic Gardens, Kew, Richmond, Surrey TW9 3AE, United Kingdom

ISBN 1 9000347 07 5

Cover photograph: *Magnolia* × *soulangeana* 'Spectabilis' by Andrew McRobb, Media Resources, Information Services Department, Royal Botanic Gardens, Kew.

Cover design by Jeff Eden for Media Resources, Information Services Department, Royal Botanic Gardens, Kew

Page make-up by Media Resources, Information Services Department, Royal Botanic Gardens, Kew, from text generated by David G. Frodin using Access 2.0® and Microsoft Word 6.0®

Printed and bound in Great Britain by Whitstable Litho Printers Ltd

Contents

Foreword

It is a pleasure to herald the start of a most useful series of publications with as a primary aim the presentation of an up-to-date global perspective of the taxonomy of selected flowering plant families. The libraries and herbaria of our botanical institutions constitute a vast resource of information on the flora of the world, but it is not as readily available to users as we might like. Sifting through the great body of published literature and multitudinous plant collections to extract and organise the basic information on plant species is a job which urgently needs to be done. David Frodin and Rafaël Govaerts have shown themselves to be adept at this, and their enviable expertise, industry and dedication to the task are now bearing fruit. The first family for which they have chosen to publish a synopsis, Magnoliaceae, is of relatively modest size, but is of considerable horticultural and public interest. It is, however, merely a prelude to other treatments already being prepared for publication, including some for major families such as the Euphorbiaceae and Rubiaceae. The prospects are exciting to those interested in a wide range of plant information. I congratulate the authors on the first of their projected series, and look forward to further and greater things to follow.

PROFESSOR SIR GHILLEAN PRANCE, F.R.S.

DIRECTOR
ROYAL BOTANIC GARDENS, KEW.
SEPTEMBER 1996

Preface

This *World Checklist and Bibliography of Magnoliaceae* is the first of a planned series intended initially to document in like fashion families and other plant groups of particular interest to the programmes of the Royal Botanic Gardens, Kew and, at the same time, to form a contribution to current international efforts to account for the world's biota.

These initiatives, among them the Species 2000 Project, the Global Plant Checklist and the Species Plantarum Project of the International Organization for Plant Information (IOPI) and the Threatened Plants Database of the World Conservation Monitoring Centre (WCMC), come at a time of a new and growing human interest in the physical and biological future of the Earth as well as significant advances in information technology. Concern in one or another form over threats to and loss of the world's biota has existed since the eighteenth century if not even before, but only from the 1960s has it become part of a serious, large-scale social and political movement (Carson, 1963; Goodland & Irwin, 1975; Wilson & Peter, 1988; Heywood & Watson, 1995). With regard to plants, the first International Union for the Conservation of Nature (IUCN) 'Red Data Book' appeared in 1970 and since then many similar works of national or more local scope have been produced. An increasing number of countries have enacted threatened plants legislation, while at an international level major developments have been the Convention on International Trade in Endangered Species of World Fauna and Flora (CITES) and the Convention on Biological Diversity (CBD), respectively effective from 1975 and 1993.

The necessity of compiling lists of threatened species naturally also led to a renewal of interest in national, state and regional biotic checklists as well as the formation or reinvigoration of biological surveys in different parts of the world. Recent 'grandes listes' for plants include *A Synonymized Checklist of the Vascular Flora of the United States, Canada and Greenland* by J. T. & R. Kartesz (1980; 2nd edn., 1994, by J. T. Kartesz alone), *Plantae Vasculares URSS* and its successor *Vascular Plants of Russia and Adjacent States* by S. K. Czerepanov (1981, 1995), *Plants of Southern Africa: Names and Distribution* by T. Arnold & B. C. de Wet (1993; previous editions by G. E. Gibbs Russell, 1984, 1985-87 as *List of Species of Southern African Plants*), *Census of Australian Vascular Plants* by R. J. Hnatiuk (1990), *Énumération des Plantes à Fleurs d'Afrique Tropicale* by J.-P. Lebrun & A. Stork (1991-), *Med-Checklist* by W. Greuter, H. M. Burdet & G. Long (1984-) and *Catalogue of the Flowering Plants and Gymnosperms of Peru* by L. Brako & J. L. Zarucchi (1993). Many works of lesser extent have also been published with all, or most, now also maintained electronically; additional works are in preparation, notably for Madagascar and China. Contemporaneous with these, but developed only in an electronic form, was the European Floristic, Taxonomic and Biosystematic Information System (or European Documentation System), derived from *Flora Europaea* and operative from 1980 to 1985 (Heywood, 1989). There have also been projects on individual families, notably Cactaceae (Hunt, 1992) and Leguminosae (under the umbrella of the International Legume Database and Information Service (ILDIS); Zarucchi *et al.*, 1994).

There has also been, not surprisingly, a growing clamour for lists of the biota of the world. To this end, following meetings at Delphi (Greece) and at Kew in 1990, interested botanists and information specialists formed a Global Species Information System (GPSIS) Action Group; the following year, it was re-established as the International Organization for Plant Information (IOPI) (Burdet, 1992; Burnett, 1994). A primary objective of this body was creation of a taxonomically validated Global Plant Checklist, a goal not successfully accomplished since publication by Kurt Sprengel of the 16th (and final) edition of Linnaeus's *Systema Vegetabilium* (1824-28) – the last 'world flora' – and the two great nineteenth-century nomenclators, *Nomenclator Botanicus* by Ernst Gottlieb von Steudel (1821-24; 2nd edn., 1840-41) and its successor, *Index Kewensis* (1893-95 and the first two supplements respectively of 1902-06 and 1904-05), compilation of which began in 1882 (Meikle, 1971).[1] The Species Plantarum Project, an initiative to produce a new World Flora, also came into being in the early 1990s and later was adopted by IOPI.

Another approach, making use of the growth and spread of the Internet and the World Wide Web, is represented by the Species 2000 Project (Bisby & Smith, 1996). Launched in 1994 under the aëgis of the International Union of Biological Sciences, Species 2000 seeks to develop access to information on all known species through electronic links with individual taxonomic and regional 'master species databases' (Bisby, 1994). A more personal contribution is the alphabetically arranged *World Checklist of Seed Plants* (Govaerts, 1995-); this so far covers the letter 'A' with 'B' expected late in October 1996 (together covering some 60 000 accepted taxa).

While the current interest in checklists bears some relationship to the conservation and biodiversity movements, such enumerative but non-descriptive works have existed since the sixteenth century. Early local floras, beginning with Johannes Thal's *Sylva Hercynia* (1588), usually were alphabetically arranged, more or less annotated catalogues; only from the eighteenth century did systematic arrangements gain greater currency. Also enumerative in character were grander works such as Caspar Bauhin's *Pinax Theatri Botanici* (1623) and John Ray's *Stirpium Europaearum extra Britannias Nascentium Sylloge* (1694). The more comprehensive works would by the end of the seventeenth century also serve as vehicles for the presentation of systems of classification, as in Ray's *Historia Plantarum* (1686-88, 1704), the botanical parts of the 10th and 12th editions of Linnaeus's *Systema Naturae* (1759, 1767) and its successor, the already-mentioned *Systema Vegetabilium* (1774 and later editions).

Taxonomically validated general nomenclators or indices did not make an appearance while the number of known plants remained more or less within human comprehension. In 1759, however, Carl Clerck produced *Nomenclator Extemporaneus Rerum Naturalium* as an index to Linnaeus's *Systema Naturae*, and from the 1770s the genre came to be seen as essential. Examples include *Nomenclator Botanicus* by Anders Retzius (1772; 2nd edn., 1782; 3rd edn. by E. A. Räuschel, 1797), originally published as an index to Linnaeus's works on plants, and L. V. F. Henckel von Donnersmarck's *Nomenclator Botanicus* (1803-12), written in conjunction with Carl Willdenow's edition of *Species Plantarum*. These works were all in the first instance systematically arranged; but from the publication of W. B. Coyte's *Index Plantarum* (1807), also written to accompany Willdenow's *Species Plantarum*, and *Nomenclator Botanicus* by A. W. Dennstedt (1810, covering 2305 genera and 20 938 species) alphabetical arrangements of taxa have prevailed. After 1815 the rapid increase in published names made one-volume works impossible and few were prepared or able to take on the task of compilation. The last to cover all plants (and fungi) was the already-mentioned first edition of Steudel's *Nomenclator Botanicus*, covering 3376 genera and 39 684 species. Its successors, likewise already noted, were the second edition of Steudel's work, covering 6722 genera and 78 005 species of seed plants alone, and *Index Kewensis*. Related indices include, for infrageneric taxa and above, *Nomenclator Botanicus* by Ludwig K. G. Pfeiffer (1871-75) and, for ferns, *Index Filicum* (1905-06), initially compiled by Carl Christensen. Until the beginning of the twentieth century, however, the major nomenclators continued to differentiate between accepted names and synonyms.[2]

In contrast to nomenclators and indices, annotated species checklists after the early nineteenth century became oriented to states, countries or regions. Significant large-scale works published before World War I include *Sylloge Florae Europaeae* (1854-55) and *Conspectus Florae Europaeae* (1878-90) by Carl Nyman, *Systematic Census of Australian Plants* (1882; 2nd edn., 1889) by Ferdinand von Mueller, *Catalogue of North American Plants North of Mexico* (1898; 2nd edn., 1900; 3rd edn. (not completed), 1909-14) by A. A. Heller, the botanical part of *Biologia Centrali-Americana* (1879-88) by William B. Hemsley, and *Index Florae Sinensis* (1886-1905), also by Hemsley with F. B. Forbes. Among those appearing after 1918 were *Enumeration of Philippine Flowering Plants* (1923-26) by Elmer D. Merrill and *Prodromus Florae Peninsulae Balcanicae* (1924-33) by August von Hayek. However, two world wars and political stalemate, economic and social dislocation, and changes in the position of systematics would all militate against truly international ventures; added to this were technological limitations and the increasing scattering of botanical information.

Worldwide surveys of genera and families did, however, continue to appear. The outstanding undertakings of the latter part of the nineteenth century were undoubtedly *Genera Plantarum* (1862-83) by George Bentham and J. D. Hooker and the original edition of

Die natürlichen Pflanzenfamilien edited by Adolf Engler and Karl Prantl (1887-1915 including supplements, with phanerogams complete by 1900). Each was complemented by a systematic index, respectively *Index Generum Phanerogamarum* (1887-88) by Théophile Durand and *Genera Siphonogamarum* (1900-07) by Karel Wilhelm von Dalla Torre and Hermann Harms. Neither, however, enjoyed a revision; and apart from *A Dictionary of the Flowering Plants and Ferns* by John C. Willis (1896-97; 6th edn., 1931; 7th edn., 1966; 8th edn., 1973) and *The Plant-book* by David Mabberley (1987) no new major indices to plant genera would appear for more than six decades. Even then, nothing quite comparable to *Genera Siphonogamarum*, with as a particular feature its coverage of principal infrageneric taxa, has yet made its appearance.

A major breakthrough came in the third quarter of the twentieth century with the introduction of effective information-processing systems. These began seriously to be exploited from the latter part of the 1960s, eventually making possible the checklists referred to at the beginning of this Introduction as well as others of lesser scope. They have also facilitated the preparation and production of such key works on higher-level taxa as *Index Nominum Genericorum (Plantarum)* (1979; supplement 1986) by E. R. Farr *et al.*, *Vascular Plant Families and Genera* (1992) by R. K. Brummitt, *Families and Genera of Spermatophytes Recognized by the Agricultural Research Service* (1992) by C. R. Gunn *et al.*, and *Names in Current Use for Extant Plant Genera* (1993) by W. Greuter *et al.* Information technology has also been essential for the present work.

D. G. FRODIN

ROYAL BOTANIC GARDENS, KEW
SEPTEMBER 1996

Notes

[1] Even then, the second edition of Steudel's *Nomenclator* was, and *Index Kewensis* has been, limited to phanerogams or seed plants. Taxonomic validation for *Index Kewensis* furthermore was effectively discontinued after 1900 (1910 for genera); ever since, it has been purely an index to published names.

[2] Mention should also be made here of the *Gray Herbarium Index*, established in 1894 in card format for published names in the Americas from 1886 onwards and accounting for infraspecific taxa (a policy not adopted by *Index Kewensis* until its sixteenth supplement (for 1971-75)).

Introduction

Aims and scope

The work presented here comprises a checklist of species and infraspecific taxa, with synonymy and indication of distribution and habit, of the Magnoliaceae of the world together with selected bibliographic references on the family including individual coverage for every accepted genus. It is the first of a planned series initiated in 1994 with a primary goal of covering families (and orders) of particular interest to the programmes of the Royal Botanic Gardens, Kew and, at the same time, to lend support to the work of the International Organization for Plant Information (IOPI) and Species 2000. Materials from a related bibliographic project begun in 1993 have also been incorporated.

Compilation of the Magnoliaceae was carried out and revised in 1995-96. There are 73 bibliographic entries while in the checklist are names of 278 accepted taxa and 724 synonyms at generic rank and below (with altogether 34 being generic names). 7 genera with 223 species are recognised. David Frodin is responsible for bibliography and commentary and Rafaël Govaerts for the checklist. Advice has been given by Hans Nooteboom (Leiden), Stephen Spongberg (Cambridge, Mass.), and others interested in the family, including participants at the international symposium 'Magnolias and their allies' at Egham (Surrey, United Kingdom) on the 12th and 13th of April 1996.

Bibliography

The bibliography is selective rather than exhaustive. For each genus, where available, only the more important references are given: monographs, complete or partial revisions, synopses and other relatively significant contributions. A selection of 'plant portraits' is also listed, the main sources being *Botanical Magazine*, *Icones Plantarum* and *Flowering Plants of Africa*. As far as possible, all references have been annotated with regard to their contents and particular features. The language(s) of each paper are given in abbreviated form. Cross-references may sometimes be given, either to papers listed under another genus or under tribes or geographical regions. Contributions on the family as a whole, along with those on particular geographical regions (where all or a part of the family is covered) and individual tribes, are listed in a block preceding the first genus. Papers thought to be of particular significance or are relatively inclusive are bulleted (•). Compilation has been carried out with the use of R:base® version 4.5, a PC database program for personal computers from Abacus/Microrim (available in MS-DOS, OS/2 and Windows formats); however, a move to Microsoft Access® is currently under consideration.

Checklist

The checklist is based on a database encompassing 24 fields and complying with the data standards proposed by IOPI, several of them in association with the Taxonomic Databases Working Group (TDWG). Compilation of the database was effected using Foxbase®, a Dbase-class database program for personal computers.

Names

Names of accepted genera and their species and infraspecific taxa are listed alphabetically. Synonymous genera (and species) are intercalated. For each accepted taxon, associated synonyms are listed chronologically if heterotypic, with any homotypic synonyms following in a given lead; in addition, all synonyms in an accepted genus are listed alphabetically at the end of that genus. Doubtful and excluded taxa are likewise summarised in an appendix following the last genus.[1] Place and date of publication of all names is given. Citation of authors follows *Authors of Plant Names* (Brummitt & Powell, 1992); for book abbreviations, the standard is *Taxonomic Literature*, 2nd edn. (Stafleu & Cowan, 1976-88); and periodicals are abbreviated according to *Botanico-Periodicum Huntianum, Supplement* (Hunt Institute of Botanical Documentation, 1991). Distributions of taxa are furnished in two ways: firstly by a generalised statement in narrative form, and secondly as TDWG geographical codes (Hollis & Brummitt, 1992) expressed to that system's third level. With respect to the latter, occurrences based on naturalisation or introduction are given in lower case. The life-form codes are based on the Raunkiær classification (1934; especially chapters 1 and 2) with modifications as outlined in *Nouvelle Flore de la Belgique, du Grand-Duché de Luxembourg, du Nord de la France et des régions voisines* (Lambinon *et al.*, 1992, pp. xx-xxi; for details, see below, pp. 6-7). A question mark (?) following a name and author indicates that a place of publication has yet to be established. Names of nothospecies ('hybrids') are preceded by a multiplication sign (×), with the place of publication being followed by the names of the parents if known. Basionyms of accepted names are designated by an asterisk (*). For genera, the number of accepted species and the geographical distribution are furnished together with general comments and the supragineric taxa as used at Kew. Type species are not indicated, but reference may be made to Farr *et al.* (1979, 1986) or Greuter *et al.* (1993).

Acceptance of species and infraspecific taxa is based on an assessment of literature with occasional reference to the herbarium where necessary. In Magnoliaceae, immediate issues are the number of genera to be accepted and, in *Magnolia*, the status of *MM. dealbata* vs. *heptapeta* and *liliiflora* vs. *quinquepeta*. With respect to genera we have chosen to follow Nooteboom (1993 and personal communication). *Talauma*, *Dugandiodendron* and *Sinomanglietia* are therefore treated as part of *Magnolia* and, where possible, the required new combinations made both in the text [and, formally, in a second appendix following the list of doubtful and excluded taxa]. Similarly, the additional segregates recognised by Liu, Xia and Yang (1995) are not accepted. As for the species names referred to above, our choice is based on a majority opinion expressed in a poll of members of the Magnolia Society and International Dendrology Society taken at the already-mentioned Egham symposium; necessary cases for conservation or rejection will be made elsewhere.

Geographical Distribution

Generalized geographical distribution is given in narrative form with everything below the third level (bar a few exceptions) of the TDWG codes given in brackets. Examples include:

SE. U.S.A. to Texas
C. America
Colombia (Antioquia)
Guatemala
China (Henan, Hubei, Guizhou, Sichuan, Yunnan)
Sumatra to Bismarck Archip.
Philippines (Palawan: Mt. Pulgar)
Cult.

For genera, distributions are furnished in the same manner, any special features being given within brackets. When the presence of a taxon in a given region or location is not certainly known, a question mark is used, e.g. New Ireland ?; when an exact location is not known, a question mark within brackets is used, e.g. Mexico (?).

Country-by-country distribution is also expressed in the form of TDWG codes (Hollis & Brummitt, 1992) taken to the third level. 'ALL' is used if the species is known to occur in every area. If the country code is not known, '+' is used. Naturalisation [not reported in Magnoliaceae] is expressed by putting the third-level codes in lower case and, if in a second-level region all occurrences are the result of naturalisation, the code number for the region is placed in brackets. The application of question marks is as indicated above for geographical regions.

Examples include:

40 ASS	[Indian subcontinent: in Assam]
42 BOR SUM	[Malesia: in Borneo, Sumatra]
75 ALL(nt. NWJ) 78 FLA	[all of Northeastern USA except New Jersey and in Southeastern USA: Florida]
79 +	[Mexico: exact distribution not known]
84 BZC BZL	[Brazil: in Brazil West-Central, Brazil Southeast]
00 CUL	[only known in cultivation]

The *region* is indicated by the two-digit number, the first digit also indicating the continent; for example, 40 is the code for the Indian Subcontinent, with the digit 4 being common to all codes for tropical Asia. The letter codes following the digits, when given, indicate 'countries' as used in the TDWG system, and usually are the first three letters of the country or other polity name or have some other mnemonic significance; for example, 'ASS' is the code for Assam, a part of India. The 'countries' are based on political units or islands or island groups.

Life-forms

The terminology for *life-forms*, definitions of which follow, is based on the system of Raunkiær (1934) with additional categories as necessary.

Main categories

Phanerophyte (phan.)
Stems: woody and indefinitely persistent
Buds: normally 3 m or more above ground
e.g.: very large shrubs; small, medium and large trees

Nanophanerophyte (nanophan.)
Stems: woody and indefinitely persistent
Buds: above soil level but normally less than 3 m above ground
e.g. shrubs such as *Magnolia stellata, Michelia figo*

Herbaceous phanerophyte (herb. phan.)
Stems: herbaceous and persisting for several years
Buds: above soil level
e.g. bananas and plantains such as *Musa basjoo*

Chamaephyte (cham.)
Stems: herbaceous and/or woody and persistent
Buds: on or just above soil level but never above 0.50 m from ground
e.g. Acaena, Alyssum, Acantholimon, Saxifraga

Hemicryptophyte (hemicr.)
Stems: herbaceous, often dying back after the growing season but with buds or growth at soil level surviving
Buds: just on or below soil level
e.g. Panax (ginseng), *Aster, Viola odorata*

Geophyte [not abbreviated]
Hemicryptophytes which survive unfavouable seasons in the form of a rhizome, bulb, tuber or root bud.
Buds: below soil level
e.g. taro, yams, lilies, tulips

Therophyte (ther.)
Plants surviving unfavourable seasons in the form of seeds, completing their life-cycle during the favourable season.
e.g. annuals (and many desert plants)

Aquatic plants

Helophyte (hel.)
Hemicryptophytes growing in soil saturated with water or in water and with leaf- and flower-bearing shoots held above water
Buds: on or below soil level
e.g. Typha, Echinodorus, Spartina, Oryza

Hydrophyte [inclusive of the three following subcategories]
Plants with stems and vegetative shoots entirely in water, the leaves usually submerged and/or floating; flower-bearing parts may emerge above the water
Buds: permanently or temporarily on the bottom of the water

Hydrohemicryptophyte (hydrohemicr.): aquatic hemicryptophytes
e.g. Stratiotes

Hydrogeophyte: aquatic geophytes [not abbreviated]
e.g. Nymphaea, Nuphar, Nymphoides

Hydrotherophyte (hydrother.): aquatic therophytes
e.g. Lemna; Utricularia vulgaris

Others

Bamboo
Biennial

Supplementary information

Climbing plants (cl.), including:
Climbing phanerophytes (cl. phan.); *e.g. Campsis radicans*
Climbing nanophanerophytes (cl. nanophan.); *e.g. Clematis florida*
Climbing hemicryptophytes (cl. hemicr.); *e.g. Vicia cracca*
Climbing tuberous geophytes (cl. tuber geophyte); *e.g. Tropaeolum tuberosum*
Climbing therophytes (cl. ther.); *e.g. Pisum sativum*
Sometimes climbing nanophanerophytes ((cl.) nanophan.)

Succulent plants (succ.), including:
Succulent nanophanerophytes (succ. nanophan.); *e.g. Opuntia ficus-indica*
Succulent chamaephytes (succ. cham.); *e.g. Lophophora williamsii*
Succulent therophytes (succ. ther.); *e.g. Sedum azureum*
Climbing succulent nanophanerophytes (cl. succ. nanophan.); *e.g. Cissus quadrangula.*

Parasitic plants (par.), including:
Hemiparasitic plants (hemipar.), i.e. parasitic plants with a continued ability to photosynthesize; *e.g. Viscum orientale* (hemipar. nanophan.)
Holoparasitic plants (holopar.), i.e. parasitic plants fully dependent upon their host; *e.g. Orobranche ramosa* (holopar. ther.), *Neottia nidus-avis* (holopar. rhizome geophyte)

Report Generation

Consolidation of output from the two databases and generation of raw reports with formatting instructions is accomplished with the use of Microsoft Access® version 2.0. Using a 'generic/text only' printer driver, the reports are then printed to a disk file. Final processing is carried out in Microsoft Word® version 6.0 with the aid of specially written macros. Both packages operate under Microsoft Windows® 3.1.

The report will also be available in Adobe Acrobat® format which is searchable and moreover compatible with a variety of existing computer platforms. There will also be the potential to include additional layers of information in any electronic release of this publication.

Acknowledgments

We wish to acknowledge with thanks the help of our advisers as well as other Kew staff, particularly R.K. Brummitt (Herbarium), S. Hinchcliffe (Computer Services) and J. Eden and M. Svanderlik (Media Resources).

Notes

[1] Many if not most of these names have never been re-evaluated since their publication and their disposition cannot yet be established.

References

Bisby, F. A. (1994). Global master species databases and biodiversity. *Biology International,* **29**: 33-40.

Bisby, F. A. & Smith, P. (1996). *Species 2000 Project Plan (version 3.1).* [2], 42 pp. Southampton: Species 2000 Secretariat, University of Southampton.

Burdet, H. M. (1992). IOPI News. *Taxon,* **41**: 390-392.

Burnett, J. (1994). IOPI and the Global Plant Checklist project. *Biology International,* **29**: 40-44.

Brummitt, R. K. (1992). *Vascular Plant Families and Genera.* 804 pp. Kew: Royal Botanic Gardens.

Brummitt, R. K. & Powell, C. E. (1992). *Authors of Plant Names.* 732 pp. Kew: Royal Botanic Gardens.

Carson, R. (1963). *Silent Spring.* xxii, 304 pp. Hamilton, London.

Farr, E. R.. Leussink, J. A. & Stafleu, F. A. (eds), 1979. *Index Nominum Genericorum (Plantarum).* 3 vols. Bohn, Scheltema & Holkema, Utrecht. (Regnum Vegetabile, 100-102).

Farr, E. R., Leussink, J. A. & Zijlstra, G. (eds), 1986. *Index Nominum Genericorum (Plantarum): Supplementum I.* xv, 126 pp. Bohn, Scheltema & Holkema, The Hague. (Regnum Vegetabile, 113.)

Goodland, R. J. A. & Irwin, H. S. (1995). *Amazon Jungle: Green Hell to Red Desert?* 161 pp., maps. Elsevier, Amsterdam.

Greuter, W. *et al.* (1993). *Names in Current Use for Extant Plant Genera* (Names in current use, 3). xxvii, 1464 pp. Koeltz, Koenigstein. (Regnum Vegetabile.)

Gunn, C. R., Wiersema, J. H., Ritchie, C. A. & Kirkbride, J. H., Jr. (1992). *Families and Genera of Spermatophytes Recognized by the Agricultural Research Service.* [10], 500 pp. United States Department of Agriculture, Washington, D.C. (Agricultural Research Service, Technical Bulletin 1796.)

Heywood, V. H. (1989). Flora Europaea and the European Documentation System. Pp. 8-10 in N. R. Morin, R. D. Whetstone, D. Wilken & K. L. Tomlinson, *Floristics for the 21st Century.* Missouri Botanical Garden, St. Louis. (Monographs in Systematic Botany from the Missouri Botanical Garden, 28.)

Heywood, V. H. (exec. ed.) & Watson, R. T. (chair) (1995). *Global Biodiversity Assessment.* xi, 1140 pp. Cambridge University Press, Cambridge (for the United Nations Environment Programme).

Hollis, S. & Brummitt, R. K. (1992). *World Geographical Scheme for Recording Plant Distributions.* ix, 105 pp. Hunt Institute for Botanical Documentation, Carnegie-Mellon University, Pittsburgh, Penna. (for the International Working Group on Taxonomic Databases for Plant Sciences). (Plant Taxonomic Database Standards, 2: version 1.0.)

Hunt, D. R. (comp.) (1992). *CITES Cactaceae Checklist.* 190 pp. Royal Botanic Gardens, Kew.

Lambinon, J. *et al.,* 1992. *Nouvelle Flore de la Belgique, du Grand-Duché de Luxembourg, du Nord de la France et des Régions Voisines.* 4th edn. cxx, 1092 pp., illus., map. Éditions du Patrimoine, Jardin Botanique National de Belgique, Meise.

Liu Yuhu, Xia Nianhe & Yang Huiqu (1995). (The origin, evolution and phytogeography of Magnoliaceae.) *J. Trop. Subtrop. Bot.* 3(4): 1-12.

Meikle, R. D. (1971). The history of the *Index Kewensis. Biol. J. Linn. Soc.,* **3**: 295-299.

Nooteboom, H. (1993). Magnoliaceae. Pp. 391-401 in K. Kubitzki (ed.), *Families and Genera of Vascular Plants,* vol. 2. Springer, Berlin.

Raunkiær, C. (1934). *The Life Forms of Plants and Statistical Plant Geography.* xvi, 632 pp., illus. Oxford University Press, London.

Wilson, E. O. (ed.) & Peter, F. M. (assoc. ed.), 1988. *BioDiversity.* xiii, 521 pp. National Academy Press, Washington, D.C.

Zarucchi, J. L., Winfield, P. J., Polhill, R. M., Hollis, S., Bisby, F. A., & Allkin, R., 1994. The ILDIS Project on the world's legume species diversity. Pp. 131-144 in Bisby, F. A., Russell, G. F., & Pankhurst, R. J., *Designs for a Global Plant Species Information System.* Oxford University Press, Oxford. (Systematics Association Special Volume, 48.)

Abbreviations

al.	alii: others
archip.	archipelago
auct.	auctoris: of author
C.	central
Ch.	Chinese
cham.	chamaephyte [life-form]
cit.	citations; cited
cl.	climber
cl. phan.	climbing phanerophyte [life-form]
cl. hemicr.	climbing hemicryptophyte [life-form]
cl. nanophan.	climbing nanophanerophyte [life-form]
cl. ther.	climbing therophyte [life-form]
cl. tuber geophyte	climbing tuberous geophyte [life-form]
Co.	comitas: county or (China) *hsien*
comb.	combinatio: combination
cons.	conservandus: conserved
cult.	cultus: cultivated
cv.	cultivarietas: cultivar
descr.	descriptio: description
Distr.	District
Du.	Dutch
E.	east(ern)
En.	English
etc.	et cetera: and the rest
e.g.	exempli gratia: for example
Fr.	French
Ge	German
hel.	helophyte [life-form]
hemicr.	hemicryptophyte
hemipar.	hemiparasitic
herb.	herbaceous
herb. phan.	herbaceous phanerophyte
holopar.	holoparasitic
hort.	hortorum: of gardens *or*
	hortulanorum: of horticulturalists
hydrohemicr.	hydrohemicryptophyte
hydrother.	hydrotherophyte
I./Is.	island/islands
ICBN	International Code of Botanical Nomenclature
i.e.	id est: that is
ign.	ignotus: unknown
in litt.	in litteris: in correspondence
ined.	ineditus: unpublished
inq.	inquilinus: naturalised
i.q.	idem quod: the same as
La.	Latin

Medit.	Mediterranean
Mt./Mts.	mountain/mountains
N.	north(ern)
nanophan.	nanophanerophyte [life-form]
nom. cons.	nomen conservandum: conserved name [ICBN]
nom. illeg.	nomen illegitimum: illegitimate name [ICBN]
nom. inval.	nomen invalidum: invalidly published name [ICBN]
nom. nud.	nomen nudum: name without a description [ICBN]
nom. rejic.	nomen rejiciendum: rejected name [ICBN]
nom. superfl.	nomen superfluum: name superfluous when published [ICBN]
nov.	novus: new
par.	parasitic
Pen.	peninsula(r)
phan.	phanerophyte (life-form)
p.p.	pro parte: partly
prov.	province
q.e.	quod est: which is
q.v.	quod vide: which see
reg.	regio: region
Rep.	republic
Ru.	Russian
S.	south(ern)
seq.	sequens: following
s.l.	sensu lato: in the broad sense
Sp.	Spanish (in literature references)
sp.	species
sphalm.	sphalmate: by mistake
s.s.	sensu stricto
st.	status
subtrop.	subtropical
syn.	synonymon: synonym
temp.	temperate
ther.	therophyte [life-form]
trop.	tropical
viz:	videlicet: namely
W.	west(ern)
?	not known, doubtful (all contexts)
*	basionym or replaced synonym (before a name)
×	nothotaxon (before a genus or species name)
+	range more than as indicated but not certainly known

Magnoliaceae

7 genera, 223 species. Americas (with patchy distribution; now absent from western North America) and from East and South Asia through Malesia to New Guinea, New Britain and (?)New Ireland. To the 7 genera recognized by Nooteboom (1993; see **General** below) and accepted here was added *Sinomanglietia* in 1994. Current standard treatments of the family appear in Nooteboom (loc. cit.) and in V. H. Heywood (ed.), *Flowering plants of the world* (1978). The primary infrageneric taxa accepted by Nooteboom (loc. cit.) for Recent members of the family are subfamilies Liriodendroideae, with a single tribe Liriodendreae, and Magnolioideae, with two tribes: Magnolieae and Michelieae. While the tribes seem clear-cut, generic limits in both Magnolieae and Michelieae have been subjects of controversy; within the Magnolieae in particular there remains some support for retention of *Dugandiodendron* and *Talauma* as segregates from *Magnolia*. There is only local interest now in recognition of more segregates in Asia. *Sinomanglietia* is not considered here to be sufficiently distinct from *Magnolia* but a more definite placement is not at this time attempted.

The family has enjoyed its present circumscription since the mid-1920s. Earlier workers, including Baillon, Prantl and Parmentier, adhered to a broader view which included the modern Schisandraceae, Illiciaceae and Winteraceae. Characteristic features include the stipules (and peduncular bracts) which in time fall and leave an annular scar around each node, spirally arranged leaves, floral parts 6 or more and usually conspicuous, monosulcate pollen, beetle pollination, numerous stamens, and carpels free to more or less concrescent (rarely syncarpous) and usually arrayed on an elongate receptacular axis. The fruiting carpels open along dorsal or ventral sutures but are sometimes circumscissile; rarely are they indehiscent (and then samaroid). There are one or more seeds in each carpel; where the latter are dehiscent the seeds, with an arilloid testa, are more or less emergent or suspended on ripening. The wood has vessels which distinguish it from that of the vesselless Winteraceae.

The primary suprageneric taxa accepted by Nooteboom (loc. cit.) for Recent members of the family are Liriodendroideae, with a single tribe Liriodendreae, and Magnolioideae, with two tribes: Magnolieae and Michelieae. These are also the highest taxa recognised by Law (Liu) (1984, 1995). In contrast, Dandy (1964, in vol. 1 of *The genera of flowering plants* by J. Hutchinson) adopted a flatter hierarchy, with two tribes: Liriodendreae and Magnolieae.

All students are agreed on the distinctness of Liriodendroideae (and *Liriodendron*), with the distinctive leaves, more 'regular' perianth, indehiscent samaroid fruiting carpels and seeds without an aril. The Magnolioideae encompasses the remaining genera; but as the group as a whole is very natural in most respects there conversely has been considerable disagreement over generic limits.

Parmentier adopted a very wide view, including all of the present Magnolidoideae in *Magnolia*. An increasing understanding of vegetative and reproductive characters, along with new discoveries, led J. E. Dandy in 1927 to a set of proposals which largely remained standard for some sixty years. In the 1980s Law and Nooteboom gave more weight to the mode of shoot growth, according this character tribal rank and so separating *Michelia* and *Elmerrillia* (Michelieae) from other genera (Magnolieae). It remains arguable, however, if the change from sympodial to monopodial shoot growth and flowers arrayed on brachyblasts represents an autapomorphy or whether it is simply an evolutionary grade and therefore a homoplasy. Other characters used to subdivide Magnolieae include:

Flowers: bisexual/unisexual
Gynoecium: sessile/stipitate
Ovules/carpel 2-4/4 or more
Mature carpels apocarpous/capsular
    longitudinally dehiscent/circumscissle
    free/more or less concrescent

The two last-named characters have attracted the most controversy, both in the michelioid and magnolioid lines. Concrescence has been in particular involved in the recognition of *Paramichelia* Hu and *Tsoongiodendron* Chun in Michelieae, and *Sinomanglietia* and *Dugandiodendron* in Magnolieae. *Talauma* was much earlier given separate recognition on account of concrescence and circumscissile fruiting carpels, while characteristic of *Aromadendron* were the numerous (18 or more) tepals and a highly concrescent fruiting body. Keng and Nooteboom have argued, however, that there are too many transitions in most characters for these additional segregates to be maintained; moreover the bicentric distribution of an otherwise weakly distinguished *Talauma* indicated that the genus might be artificial. Similarly, Baranova (1990; see **Magnolia**) on stomatographical evidence suggested that *Dugandiodendron* was too slightly founded to be maintained. Wood anatomy in general shows a similar pattern of weak distinctions within characters.

More recently, results from molecular systematic studies in *Magnolia* sect. *Rytidospermum* led Qiu, Chase and Parks (1995) to suggest that *all* past characters needed re-evaluation, given that the section was found to be embedded in a cladogram inclusive of representatives from the entire subfamily. This lends weight to the argument of Nooteboom (International Dendrology Society, 1996: 4) that the two centres of distribution of the former *Talauma* represent species respectively more closely related to *Magnolia* in those areas than to each other. A more extensive molecular systematic study of the family is awaited.

Bibliography

No separate bibliography of the family has been published. Useful lists are contained in Spongberg (1976) and Chen & Nooteboom (1993; for both, see below). For special aspects the references in Nooteboom (in Kubitzki 1993: 400-401; see **General**) may be consulted.

General

Dandy (1927, with later revisions including one in 1964 for the first volume of *The Genera of Flowering Plants* by J. Hutchinson) has long served as a basic survey of genera, but is now out of date. At present 7 are generally accepted (cf. Nooteboom, 1993). Law (Liu) (1984, 1995) accepts 15, exclusive of the recently published *Sinomanglietia* (which here is reduced to *Magnolia*).

- de Candolle, A.-P. (1817). Magnolieae. Regni vegetabilis systema naturae 1: 449-560. Paris: Treuttel & Würtz. La. — First (and last, apart from the author's 1824 adaptation for his *Prodromus*) complete treatment of the family.

  Baillon, H. (1866). Mémoire sur la famille des Magnoliacées. Adansonia 7: 1-16. Fr. — Commentary on characters and generic delimitation (with a broader rather than narrower view). Covers also some genera now in 'allied' families, with among the latter Winteraceae and Canellaceae.

  Parmentier, P. (1895). Histoire des Magnoliacées. Bull. Sci. France Belgique 27: 159-331. Fr. — Monographic, with emphasis on leaf and stem anatomy; includes extensive general discussion followed by descriptive systematic treatments, without keys, of 8 genera (only two of which, *Magnolia* s. latiss. and *Liriodendron* with 58 species in all, are in Magnoliaceae as currently delimited) with emphasis on anatomical features. Now largely of historical interest.

- Dandy, J. E. (1927). The genera of Magnolieae. Bull. Misc. Inform. Kew (1927): 257-264. En. — Generic revision, with key. Precursor to a projected (but never completed) revision of the family.

  Law, Y. W. (Liu Yuhu) (1984). (A preliminary study on the taxonomy of the family Magnoliaceae.) Acta Phytotax. Sin. 22: 89-109. Ch. — A new system of the family is proposed, with recognition of 15 genera (14 in Magnolioideae, disposed in 2 tribes (each with 2 subtribes), and 1 in Liriodendroideae, their distribution being shown on the map). The arrangement of taxa in the synopsis (pp. 105-106) is unfortunately confusing.

- Nooteboom, H. P. (1985). Notes on Magnoliaceae: with a revision of *Pachylarnax* and *Elmerrillia* and the Malesian species of *Manglietia* and *Michelia*. Blumea 31(1): 65-121. En. — Includes revisions of *Magnolia* (supraspecific level only), *Manglietia* (Malesian species), *Pachylarnax*, *Kmeria*, *Michelia* (Malesian species) and *Elmerrillia*; also notes on generic delimitation, including the disposition of *Dugandiodendron*. In part precursory to a family treatment in *Flora Malesiana* (see **Malesia**).
- Nooteboom, H. P. (1987). Notes on Magnoliaceae, II. See ***Magnolia***.
- Chen, B.-L. & H. P. Nooteboom (1993). Notes on Magnoliaceae, III. See **Asia**.
  Nooteboom, H. (1993). Magnoliaceae. In K. Kubitzki (ed.), The families and genera of vascular plants, 2: 391-401. Berlin. — Includes an extended treatment of the family and keys to and descriptions of the genera aong with their distribution, chromosome numbers, number of species and brief notes on taxonomy and special features.
  Liu Yuhu, Xia Nianhe & Yang Huiqu (1995). (The origin, evolution and phytogeography of Magnoliaceae.) J. Trop. Subtrop. Bot. 3(4): 1-12, illus., maps. Ch. — A revised system of the family is proposed, and former systems compared. The 15 genera of Law (1984) are all retained; at suprageneric level a further subtribe is added. Evolutionary tree on p. 9 and maps of individual genera on pp. 10-11.

Special

For the many papers published since the comparative survey of Canright (1952-60), see Nooteboom in Kubitzki (1993, under General References: 400-401). Canright's references will furnish a useful guide to older literature.

Good, R. d'O. (1925). The past and present distribution of the Magnolieae. Ann. Bot. 39: 409-430. En. — Phytogeographical, past and present. A precursor to Dandy & Good (1929).

Dandy, J. E. & R. d'O. Good (1929). Magnoliaceae Jaume. In E. Hannig and H. Winkler (eds), Pflanzenareale 2(5): 35-38, maps 41-43 (each with 2 maps). Jena: Fischer. Ge. — Discussion of distribution, genus by genus; the area of each genus mapped. The patchy distribution of *Magnolia* (and the former *Talauma*) in the Americas is clearly evident from maps 41a and 42a, even given subsequent extensions (e.g. in the Guayana Highland region). See also Good (1925).

Canright, J. E. (1952-60). The comparative morphology and relationships of the Magnoliaceae, I-IV. Amer. J. Bot. 39: 484-497 (I); Phytomorphology 3: 355-365 (II, 1953); Amer. J. Bot. 47: 145-455 (III, 1960); J. Arnold Arbor. 36: 119-140 (IV, 1955). En. — Considerations of the morphology of the stamens, pollen, and carpels and of wood and nodal anatomy.

Americas

In the Americas native genera include *Liriodendron* and *Magnolia* (including *Dugandiodendron* and *Talauma*). Middle and South American family references appear under *Magnolia*. The family was formerly well represented in western North America north of Mexico, but by the late Tertiary all species had died out; currently none occurs north of southern Sonora in Mexico.

Howard, R. A. (1948). See ***Magnolia***.

Wood, C. E. (1958). The genera of the woody Ranales in the southeastern United States. J. Arnold Arbor. 39: 296-346. (Generic flora of the Southeastern United States.) En. — Includes treatments of *Magnolia* and *Liriodendron*, with references. For more detailed coverage, see Spongberg (1976).

- Spongberg, S. A. (1976). Magnoliaceae hardy in temperate North America. J. Arnold Arbor. 57: 250-311. En. — A taxonomic treatment, with keys and descriptions, of all representatives of the family in North America north of Mexico, both native and more or less extensively cultivated (whatever their origin or status); extensive references. Covers 22 normal and 4 hybrid species of *Magnolia*, 1 of *Michelia*, and all of *Liriodendron*.

Asia

Monsoonal Asia is a major centre for Magnoliaceae; 6 of the 7 genera are present. Included here are the treatments by George King for South Asia (1891) and by B.-L. Chen & H. P. Nooteboom for China (1993). The recently published *Sinomanglietia* is here referred to *Magnolia*.

- King, G. (1891). The Magnoliaceae of British India. Ann. Roy. Bot. Gard. Calcutta 3(2): 197-226, pl. 38-74. En. — Systematic treatment (in a generous format), with descriptions, citations of exsiccatae, distribution and commentary; features numerous large plates.
- Keng, H. (1975). Magnoliaceae. Fl. Thailand 2(3): 251-267, illus. En. — Flora treatment; the most recent available in SE Asia.
- Chen, B.-L. & H. P. Nooteboom (1993). Notes on Magnoliaceae, III. The Magnoliaceae of China. Ann. Missouri Bot. Gard. 80(4): 999-1104. En. — Complete treatment, with keys, covering 5 genera with 81 indigenous and 7 cultivated species; synopsis; extensive bibliography. A precursor to the English edition of *Flora reipublicae popularis sinicae*. A successor in large part to A. Rehder & E. H. Wilson, 1913. Magnoliaceae. In C. S. Sargent (ed.), Plantae Wilsonianae 1: 391-418.

Malesia

Elmerrillia (endemic), *Magnolia*, *Manglietia*, *Michelia*, and *Pachylarnax* are represented with 33 species.

- Korthals, P. (1851). Bijdrage tot de kennis der Indische Magnoliaceae. Ned. Kruidk. Arch. 2: 91-98. Du. — A first overall contribution to Malesian Magnoliaceae.
- Nooteboom, H. P. (1988). Magnoliaceae. Fl. Malesiana, I, 10(3): 561-605. Dordrecht. En. — 5 genera, 35 species; keys, descriptions, distributions, commentary, illustrations and maps with a substantial introductory part. [Some changes to this treatment are now necessary due to subsequent work by the author and others in E. and SE. Asia.]

Alcimandra

Synonyms:
Alcimandra Dandy === **Magnolia** L.
Alcimandra cathcartii (Hook.f. & Thomson) Dandy === **Magnolia cathcartii** (Hook.f. & Thomson) Noot.

Aromadendron

Synonyms:
Aromadendron Blume === **Magnolia** L.
Aromadendron baillonii (Pierre) Craib === **Michelia baillonii** (Pierre) Finet & Gagnep.
Aromadendron elegans Blume === **Magnolia elegans** (Blume) H.Keng
Aromadendron elegans var. *glauca* (Korth.) Dandy === **Magnolia elegans** (Blume) H.Keng
Aromadendron glaucum Korth. === **Magnolia elegans** (Blume) H.Keng
Aromadendron nutans Dandy === **Magnolia bintuluensis** (A.Agostini) Noot.
Aromadendron oreadum (Diels) Kaneh. & Hatus. === **Magnolia liliifera** (L.) Baill. var. **liliifera**
Aromadendron spongocarpum (King) Craib === **Michelia baillonii** (Pierre) Finet & Gagnep.
Aromadendron yunnanense Hu === ?

Blumia

Synonyms:
Blumia Nees ex Blume === **Magnolia** L.
Blumia candollei (Blume) Nees === **Magnolia liliifera** (L.) Baill. var. **liliifera**

Buergeria

Synonyms:
Buergeria Siebold & Zucc. === **Magnolia** L.
Buergeria obovata Siebold & Zucc. === **Magnolia kobus** DC.
Buergeria salicifolia Siebold & Zucc. === **Magnolia salicifolia** (Siebold & Zucc.) Maxim.
Buergeria stellata Siebold & Zucc. === **Magnolia stellata** (Siebold & Zucc.) Maxim.

Champaca

Synonyms:
Champaca Adans. === **Michelia** L.
Champaca fasciculata Noronha === ?
Champaca michelia Noronha === **Michelia champaca** L. var. **champaca**
Champaca turbinata Noronha === ?
Champaca velutina Kuntze === **Michelia champaca** var. **pubinervia** (Blume) Miq.

Dugandiodendron

Synonyms:
Dugandiodendron Lozano === **Magnolia** L.
Dugandiodendron argyrothrichum Lozano === **Magnolia argyrothricha** (Lozano) Govaerts
Dugandiodendron calimaense Lozano === **Magnolia calimaensis** (Lozano) Govaerts
Dugandiodendron calophyllum Lozano === **Magnolia calophylla** (Lozano) Govaerts
Dugandiodendron cararense Lozano === **Magnolia cararensis** (Lozano) Govaerts
Dugandiodendron chimantense (Steyerm. & Maguire) Lozano === **Magnolia chimantensis** Steyerm. & Maguire
Dugandiodendron colombianum (Little) Lozano === **Magnolia colombiana** (Little) Govaerts
Dugandiodendron guatapense Lozano === **Magnolia guatapensis** (Lozano) Govaerts
Dugandiodendron lenticellata Lozano === **Magnolia lenticellatum** (Lozano) Govaerts
Dugandiodendron magnifolium Lozano === **Magnolia magnifolia** (Lozano) Govaerts

Dugandiodendron mahechae Lozano === **Magnolia mahechae** (Lozano) Govaerts
Dugandiodendron ptaritepuianum (Steyerm.) Lozano === **Magnolia ptaritepuiana** Steyerm.
Dugandiodendron striatifolium (Little) Lozano === **Magnolia striatifolia** Little
Dugandiodendron urraoense Lozano === **Magnolia urraoense** (Lozano) Govaerts
Dugandiodendron yarumalense Lozano === **Magnolia yarumalense** (Lozano) Govaerts

Elmerrillia

4 species, Malesia (Sumatra, Borneo, Sulawesi and Philippines to New Guinea and Bismarck Archip.). (Michelieae)

- Nooteboom, H. P. (1985). Notes on Magnoliaceae. Blumea 31(1): 65-121. En. — *Elmerrillia*, pp. 100-108. Revision (4 species, one with additional varieties); key.

Elmerrillia Dandy, Bull. Misc. Inform. Kew 1927: 261 (1927).
Philippines, Sulawesi to Bismarck Archip. 42.

Elmerrillia ovalis (Miq.) Dandy, Bull. Misc. Inform. Kew 1927: 261 (1927).
Sulawesi, Moluccas. 42 MOL SUL. Phan.
* *Talauma ovalis* Miq., Ann. Mus. Bot. Lugduno-Batavi 4: 69 (1868).
Talauma vrieseana Miq., Ann. Mus. Bot. Lugduno-Batavi 4: 70 (1868). *Magnolia vrieseana* (Miq.) Baill. ex Pierre, Fl. Forest. Cochinch.: 2 (1880). *Elmerrillia vrieseana* (Miq.) Dandy, Bull. Misc. Inform. Kew 1927: 262 (1927).

Elmerrillia platyphylla (Merr.) Noot., Blumea 31: 102 (1985).
Philippines (Leyte, Mindanao, Zamboanga). 42 PHI. Phan.
* *Michelia platyphylla* Merr., Philipp. J. Sci., C 13: 11 (1918).

Elmerrillia pubescens (Merr.) Dandy, Bull. Misc. Inform. Kew 1927: 261 (1927).
Philippines (Mindanao). 42 PHI. Phan.
* *Talauma pubescens* Merr., Philipp. J. Sci. 3: 133 (1908).

Elmerrillia tsiampacca (L.) Dandy in S.Nilsson, World Pollen Spore Fl. 3(Magnoliaceae): 5 (1974).
Sumatra to Bismarck Archip. 42 BIS BOR MOL NWG SUL SUM. Phan.
* *Michelia tsiampacca* L., Mant. Pl. 1: 78 (1767).

var. **glaberrima** (Dandy) Noot., Blumea 31: 107 (1985).
New Guinea. 42 NWG. Phan.
* *Elmerrillia papuana* var. *glaberrima* Dandy, Bull. Misc. Inform. Kew 1928: 185 (1928).

subsp. **mollis** (Dandy) Noot., Blumea 31: 108 (1985).
Sumatra, Borneo. 42 BOR SUM. Phan.
* *Elmerrillia mollis* Dandy, Bull. Misc. Inform. Kew 1928: 184 (1928). *Michelia mollis* (Dandy) McLaughlin, Trop. Woods 34: 36 (1933).

subsp. **tsiampacca**
C. & N. Sulawesi, Moluccas, New Guinea, Bismarck Archip. (New Britain; New Ireland ?). 42 BIS MOL NWG SUL. Phan.
Michelia celebica Koord., Meded. Lands Plantentuin 19: 631 (1898). *Elmerrillia celebica* (Koord.) Dandy, Bull. Misc. Inform. Kew 1927: 261 (1927).
Talauma papuana Schltr., Bot. Jahrb. Syst. 50: 70 (1913). *Elmerrillia papuana* (Schltr.) Dandy, Bull. Misc. Inform. Kew 1927: 261 (1927).
Michelia forbesii Baker f., J. Bot. 61(Suppl.): 2 (1923).
Michelia arfakiana A.Agostini, Atti Reale Accad. Fisiocrit. Siena, IX, 7: 25 (1926).
Elmerrillia papuana var. *adpressa* Dandy, Bull. Misc. Inform. Kew 1928: 185 (1928).
Elmerrillia sericea C.T.White, J. Arnold Arbor. 10: 212 (1929).

Synonyms:
Elmerrillia celebica (Koord.) Dandy === **Elmerrillia tsiampacca** (L.) Dandy subsp.
  tsiampacca
Elmerrillia mollis Dandy === **Elmerrillia tsiampacca** subsp. **mollis** (Dandy) Noot.
Elmerrillia papuana (Schltr.) Dandy === **Elmerrillia tsiampacca** (L.) Dandy subsp.
  tsiampacca
Elmerrillia papuana var. *adpressa* Dandy === **Elmerrillia tsiampacca** (L.) Dandy subsp.
  tsiampacca
Elmerrillia papuana var. *glaberrima* Dandy === **Elmerrillia tsiampacca** var. **glaberrima**
  (Dandy) Noot.
Elmerrillia sericea C.T.White === **Elmerrillia tsiampacca** (L.) Dandy subsp. **tsiampacca**
Elmerrillia vrieseana (Miq.) Dandy === **Elmerrillia ovalis** (Miq.) Dandy

Guillimia

Synonyms:
Guillimia Rchb. === **Magnolia** L.

Gwillimia

Synonyms:
Gwillimia Rottler ex DC. === **Magnolia** L.
Gwillimia indica Rottler ex DC. === **Magnolia coco** (Lour.) DC.
Gwillimia yulan (Desf.) Kostel. === **Magnolia denudata** Desr.

Kmeria

2 species, SE Asia (Cambodia, Thailand), S. and SW. China. (Magnolieae)

- Nooteboom, H. P. (1985). Notes on Magnoliaceae. Blumea 31(1): 65-121. En. — *Kmeria*,
  pp. 98-99. Brief note (*K. duperreana* the sole species known south of China).

Kmeria Dandy, Bull. Misc. Inform. Kew 1927: 262 (1927).
  S. China, Indo-China. 36 41.

Kmeria duperreana (Pierre) Dandy, Bull. Misc. Inform. Kew 1927: 262 (1927).
  Cambodia, Thailand, Vietnam. 41 CBD THA VIE. Phan.
  *Magnolia duperreana Pierre, Fl. Forest. Cochinch.: 1 (1880). *Talauma duperreana* (Pierre)
    Finet & Gagnep., Bull. Soc. Bot. France 52(4): 32 (1906).

Kmeria septentrionalis Dandy, J. Bot. 69: 233 (1931).
  S. China (N. & C. Guangxi, Yunnan, Guizhou). 36 CHC CHS. Phan.

Kobus

Synonyms:
Kobus Kaempf. ex Salisb. === **Magnolia** L.
Kobus acuminata (L.) Nieuwl. === **Magnolia acuminata** (L.) L.

Lassonia

Synonyms:
Lassonia Buc'hoz === **Magnolia** L.
Lassonia heptapeta Buc'hoz === **Magnolia denudata** Desr.
Lassonia quinquepeta Buc'hoz === **Magnolia liliiflora** Desr.

Lirianthe

Synonyms:
Lirianthe Spach === **Magnolia** L.
Lirianthe grandiflora Spach === **Magnolia pterocarpa** Roxb.

Liriodendron

2 species, E North America and China (1 species each), mutually closely related (and by some older authors combined). Both species are treated together in the works of Shaparenko (1937), Spongberg (1976; see **Americas**) and Chen & Nooteboom (1993; see **Asia**). (Liriodendreae)

Bean, W. J. (1919). The two tulip trees. Gard. Chron., III, 65: 128. En. — Oriented towards arboriculture.

- Shaparenko, K. K. (1937). Tyul'pannoe derevo [On tulip-trees]. Trudy Bot. Inst. Akad. Nauk SSSR, I, 4: 93-170, illus., maps, 2 pls. Ru. — Monograph of Recent and fossil species; includes history of research and a formal treatment of genus (pp. 98-102, with descriptions of the 2 modern species and exsiccatae but no key), followed by chapters on development, historical geography and phylogeny (fossil forms listed, pp. 145-149); references and English summary at end.

Santamour, F. S., Jr. & F. G. Meyer (1971). The two tuliptrees. Amer. Hort. Mag. 1971(Spring): 87-89, illus. En. — Report of trials, particularly with *Liriodendron chinense*; includes a diagnostic photograph of leaves, flowers and samaras of the two species.

- Spongberg, S. A. (1976). Magnoliaceae hardy in temperate North America. J. Arnold Arbor. 57: 250-312. En. — *Liriodendron*, pp. 308-312. Treatment of the 2 species (1 native), with key and references.

Schoenike, R. E. (1980). Yellow-poplar (*Liriodendron tulipifera* L.): an annotated bibliography to and including 1974. xix, unnumbered pp.; pp. A1-A76, S1-S38. Clemson, S.C. En. — 5891 annotated references.

Grey-Wilson, C. (1982). *Liriodendron chinense*. Bot. Mag. 184(1): pl. 843. En. — Plant portrait; includes description and synonymy. [The tree is smaller than the better-known *L. tulipifera*; in addition, its native range is relatively limited.]

Parks, C. R., N. G. Miller, J. F. Wendel & K. M. McDougal (1983). Genetic divergence within the genus *Liriodendron* (Magnoliaceae). Ann. Missouri Bot. Gard., 70: 658-666. En. — Includes data on interspecific hybrids; a high compatibility remains despite long mutual isolation.

Parks, C. R. & J. F. Wendel (1990). Molecular divergence between Asian and North American species of *Liriodendron* (Magnoliaceae) and implications for interpretation of fossil floras. Amer. J. Bot., 77: 1243-1256. En. — Molecular markers from allozymes and plastid genomes analysed; from measures of genetic distance estimates of 10-16 and 11-14 million years B.P. were obtained, comparable to time estimates from the geological record for the effective sundering of the warm-temperate mixed forests of Asia and North America (no later than 13 million years B.P.). A relatively high degree of interspecific compatibility was shown to have been maintained since the Miocene and thus is not necessarily evidence of Pleistocene or Holocene divergence.

Liriodendron L., Sp. Pl.: 535 (1753).
  E. Canada, NC. & E. U.S.A., China. 36 72 74 75 78. Phan.
    Tulipifera Mill., Gard. Dict. Abr. ed. 4(1754).

Liriodendron chinense (Hemsl.) Sarg., Trees & Shrubs 1: 103 (1903).
  China (Anhui, Guangxi, Jiangsu, Guizhou, Hubei, Hunan, Jiangxi, Shaanxi, Zhejiang, Sichuan, Yunnan). 36 CHC CHN CHS. Phan.
    * Liriodendron tulipifera var. *chinense* Hemsl., J. Linn. Soc., Bot. 23: 29 (1886).
    Liriodendron tulipifera var. *sinensis* Diels, Bot. Jahrb. Syst. 29: 322 (1900).

Liriodendron tulipifera L., Sp. Pl.: 535 (1753). *Tulipifera liriodendrum* Mill., Gard. Dict. ed. 8(1768).

S. Ontario, NC. & E. U.S.A. 72 ONT 74 ILL MSO 75 CNT INI MAS MIC NWJ NWY OHI PEN RHO VER WVA 78 ALL. Phan.

Liriodendron procerum Salisb., Prodr. Stirp. Chap. Allerton: 379 (1796).
Liriodendron integrifolium Steud., Nomencl. Bot. ed. 1: 486 (1824).
Liriodendron heterophyllum K.Koch, Dendrologie 1: 381 (1869).
Liriodendron obtusilobum K.Koch, Dendrologie 1: 381 (1869).
Liriodendron fastigiatum Dippel, Handb. Laubholzk. 3: 155, 736 (1893).
Liriodendron truncatifolium Stokes, Bot. Mat. Med. 3: 233 (1912).

Synonyms:
Liriodendron coco Lour. === **Magnolia coco** (Lour.) DC.
Liriodendron fastigiatum Dippel === **Liriodendron tulipifera** L.
Liriodendron figo Lour. === **Michelia figo** (Lour.) Spreng.
Liriodendron grandiflorum Roxb. === **Magnolia pterocarpa** Roxb.
Liriodendron heterophyllum K.Koch === **Liriodendron tulipifera** L.
Liriodendron indicum Spreng. === **Magnolia pterocarpa** Roxb.
Liriodendron integrifolium Steud. === **Liriodendron tulipifera** L.
Liriodendron liliiferum L. === **Magnolia liliifera** (L.) Baill.
Liriodendron liliiflorum Steud. === **Magnolia obovata** Thunb.
Liriodendron obtusilobum K.Koch === **Liriodendron tulipifera** L.
Liriodendron procerum Salisb. === **Liriodendron tulipifera** L.
Liriodendron truncatifolium Stokes === **Liriodendron tulipifera** L.
Liriodendron tulipifera var. *chinense* Hemsl. === **Liriodendron chinense** (Hemsl.) Sarg.
Liriodendron tulipifera var. *sinensis* Diels === **Liriodendron chinense** (Hemsl.) Sarg.

Liriopsis

Synonyms:
Liriopsis Spach === **Michelia** L.
Liriopsis fuscata (Andrews) Spach === **Michelia figo** (Lour.) Spreng. var. **figo**
Liriopsis pumila Spach ex Baill. === **Magnolia coco** (Lour.) DC.

Magnolia

128 species, Asia and Malesia (with *Magnolia liliifera* reaching New Guinea) and in the Americas (though now absent from W North America north of Mexico); includes *Alcimandra, Aromadendron, Dugandiodendron, Manglietiastrum* and *Talauma*. No full modern survey is available, nor is there a phylogenetic classification. Existing treatises, in the main oriented towards enthusiasts, emphasise the more 'temperate' species and moreover have yet to account for the reduction of *Talauma*. Nooteboom (1985: 84-91) furnishes the most recent system of the genus. This comprises 16 sections in 3 subgenera; in comparison with Dandy's treatment in Treseder (1978: 29-36) sect. *Alcimandra* (formerly *Alcimandra* and part of *Michelia*) is added to subgen. *Magnolia* and *Talauma* appears as a third subgenus, with 3 sections (1 American, 2 Asian/Malesian) formerly in that genus (and *Aromadendron*) and 1 (in China) based on *Manglietiastrum sinicum* Y. W. Law. *Dugandiodendron* is now distributed between subgen. *Magnolia* and subgen. *Talauma*; its type species, *D. mahechae*, is probably in the former on account of a perceived relationship with *M. chimantensis* (cf. Lozano-Contreras, 1994). The recently published *Sinomanglietia* (Yu, 1994), initially considered to be related to *Manglietia*, is also here reduced to *Magnolia* but no further placement is essayed. It has recently been argued that the present infrageneric classification is not ideal, reflecting 'a lack of work on the group'; Dandy's 1978 system and its modifications remain to be seriously tested (Qiu, Chase & Parks, 1995; Qiu, Parks & Chase, 1995). A similar conclusion was reached by Tobe (1993, unpubl.). (Magnolieae).

Engler, A., F. Pax & P. Graebner (1902). Die Verbreitung wichtiger Baumgattungen kartographischer dargestellt, zum Gebrauch in Botanischen Gärten und Museum sowie bei Vorlesungen. Notizbl. Bot. Gart. Mus. Berlin 3: 181-182, maps. Ge. — Comprises a distribution map (now very out of date).

Dandy, J. E. (1927). Key to the species of *Magnolia*. J. Roy. Hort. Soc. 52: 260-264. En. — Key to 44 species (*Magnolia* sensu stricto), revised from that in *Magnolias* (1927) by J.G. Millais. Does not include the former *Talauma* or *Aromadendron*.

Cheng, W. C. (1934). The genus *Magnolia* in China. J. Bot. Soc. China 1: 280-305. Ch. — Regional revision; not cited by Johnstone (1955). Succeeded by Chen & Nooteboom (1993).

• Howard, R. A. (1948). The morphology and systematics of the West Indian Magnoliaceae. Bull. Torrey Bot. Club 75: 335-357. En. — Includes a regional revision of *Magnolia* sens. lat. (11 native and 1 cultivated species) with keys, synonymy, descriptions, indication of distribution, exsiccatae, vernacular names and commentary. [The 8 species then assigned to sect. *Theorhodon* now form sect. *Splendentes* (Vásquez-G. 1994).]

• Dandy, J. E. (1950). A survey of the genus *Magnolia* together with *Manglietia* and *Michelia*. In Royal Horticultural Society (Great Britain), Camellias and Magnolias: 64-81. London. En. — Discursive synoptic survey of the genus and its sections, species being listed in passing in the text; foldout key to subgenera and sections. There are also brief accounts of *Manglietia* and *Michelia*. [Partly succeeded by the author's 1978 survey (see below).]

• Royal Horticultural Society (Great Britain) (1950). Camellias and magnolias: report of the conference held by the Royal Horticultural Society, April 4-5, 1950 (ed. P. M. Synge). 134 pp., illus. London. En. — Comprises papers on a variety of topics, including a survey of the genus by Dandy (separately cited) and a key to East Asian species by G. H. Johnstone (pp. 44-52).

Johnstone, G. H. (1955). Asiatic magnolias in cultivation. 155 pp., illus. 14 col. pl., col. frontispiece, folding map. London: Royal Horticultural Society. En. — Includes a sectional synopsis with key (pp. 36-38) and detailed treatments of 18 species with 6 additional infraspecific taxa; index, pp. 153-154. Based in the first instance on the collection of the author in Cornwall (England). See also the author's key to E Asian species in Royal Horticultural Society, 1950. *Camellias and magnolias* (cited above).

• Felger, R. S. (1971). The distribution of *Magnolia* in northwestern Mexico. J. Arizona Acad. Sci. 6: 251-253. En. — Of particular interest given the modern absence of the genus in western North America north of the Mexican border. *M. pacifica* occurs from southern Sonora southwards.

• Spongberg, S. A. (1976). Magnoliaceae hardy in temperate North America. J. Arnold Arbor. 57: 250-312. En. — *Magnolia*, pp. 254-306. Treatment of 26 species and hybrids (including all those native), with key and references.

• Dandy, J. E. (1978). A revised survey of the genus *Magnolia* together with *Manglietia* and *Michelia*. In N. G. Treseder, *Magnolias*: 29-37. London. En. — Discursive synoptic survey of the genus and its sections, similar to that of 1950; no key. Brief accounts are given for *Manglietia* and *Michelia*. With Nooteboom's revisions (1985; see below), this classification remains standard for the genus.

• Keng, H. (1978). The delimitation of the genus *Magnolia* (Magnoliaceae). Gard. Bull. Singapore 31(2): 127-131. En. — A proposal to unite *Talauma* with *Magnolia* is here put forward (see also Nooteboom 1985), along with *Aromadendron* and *Manglietia*.

• Treseder, N. G. (1978). Magnolias. xviii, 243, [3] pp., illus. (part col.), maps. London: Faber and Faber. En. — At present the primary modern reference for enthusiasts; includes J. E. Dandy's last survey of the genus (together with some comments on *Manglietia* and *Michelia*; see below) as well as a detailed treatment of the 'temperate' species (in 2 subgenera with 9 sections) with descriptions, synonymy, illustration references, distributions, and extensive commentaries. [A successor to J. G. Millais, 1927. *Magnolias*. London.]

Hernández-Cerda, M. E. (1980). Magnoliaceae. Fl. Veracruz 14: 1-14, illus., maps. Sp. — Treatment of 4 species (3 native) in 2 genera (both now in *Magnolia*).

Treseder, N. G. (1981). The book of magnolias. 96 p., text-fig., 33 col. pl. London: Collins. En. — Coloured paintings of selected species, varieties and hybrids with descriptive facing text, followed by appendices on early records, magnolia hunters (E. H. Wilson and G. Forrest), flowers and fruits, and growing and propagation tips; glossary at end but no list of references.

- Lozano-Contreras, G. (1983). Magnoliaceae. Fl. Colombia 1: 1-119, illus., maps. Sp. — Treatment of 2 native genera (*Dugandiodendron* and *Talauma*, both now in *Magnolia*) with 24 species, as well as 2 introduced species of *Magnolia* sensu str.

Lozano-Contreras, G. (1984). Consideraciones sobre el genero *Dugandiodendron* (Magnoliaceae). Taxon 33(4): 691-696. Sp. — A defence of *Dugandiodendron*.

Nooteboom, H. P. (1984). *Dugandiodendron* (Magnoliaceae) erroneously described. Taxon 33(4): 696-698. En. — *Dugandiodendron* shown to have been based on imperfectly described differential characters.

- Nooteboom, H. P. (1985). Notes on Magnoliaceae. Blumea 31(1): 65-121. En. — *Magnolia*, pp. 83-91. A supraspecific revision (16 sections in 3 subgenera), with key modified from that of Dandy (in Treseder 1978); also includes a synopsis of sect. *Gynopodium* (one of those of Dandy). The former *Talauma* has here been ranked as the third subgenus (reviving an 1881 proposal of Louis Pierre); this incorporates *Aromadendron* and *Manglietiastrum*.

Schnetter, M. L. & G. Lozano-Contreras (1985). Contribución al conocimiento de la estructura foliar de las especies de Magnoliaceas colombianas. Caldasia 14(67): 193-206. Sp. — A systematic treatment of the leaf micromorphology of 21 species of *Magnolia* (9 and 12 respectively in the former *Dugandiodendron* and *Talauma*), based on Lozano-Contreras (1983; see above).

Ueda, K. (1985). A nomenclatural revision of the Japanese *Magnolia* species (Magnoliaceae), together with two long-cultivated Chinese species. I. *M. hypoleuca*; II. *M. tomentosa* and *M. praecocissima*. Taxon 35: 340-344, 344-347. En. — Proposals for name changes; in the second paper *M. praecocissima* Koidz. is taken up for the well-known but non-'typical' element of *M. kobus* DC. [See Nooteboom (1994) for a counter-proposal.]

- Ueda, K. (1985). A nomenclatural revision of the Japanese *Magnolia* species (Magnoliaceae), together with two long-cultivated Chinese species. III, *M. heptapeta* and *M. quinquepeta*. Acta Phytotax. Geobot. 36: 149-161. En. — Incorporates a full enumeration, with synonymy and suprageneric disposition, of Japanese (including two from China, long-cultivated) magnolias. Incorporates the results of the author's two other papers, though published ahead of them.

- Meyer, F. G. & E. McClintock (1987). Rejection of the names *Magnolia heptapeta* and *M. quinquepeta* (Magnoliaceae). Taxon 36(3): 590-600, illus. En. — The revival by recent workers of these two names of Pierre Buc'hoz, first published in *Lassonia* and transferred to *Magnolia* by Dandy in 1934, has led to considerable controversy. The authors argue that, while the Chinese names on Buc'hoz's figures are correctly applicable, the figures themselves do not adequately characterise the plants concerned. The names should accordingly be rejected as *incertae sedis* and the next available alternatives, respectively *MM. denudata* and *liliiflora*, adopted.

- Nooteboom, H. P. (1987). Notes on Magnoliaceae, II. Revision of *Magnolia* sections Maingola (Malesian species), Aromadendron and Blumiana. Blumea 32(2): 343-382. En. — Comprises a species-level revision of *Magnolia* (now including *Aromadendron* and *Talauma*, with the Malesian species in the last-named corresponding to sect. *Blumiana*). Precursory to a family treatment in *Flora Malesiana* (see **Malesia**).

Seitner, P. G. (1989). A nomenclature reference for the genus *Magnolia* with emphasis on species and hybrids of more temperate climates. Unpaged, loose-leaf. Chicago, Ill.: The author. En. — A nomenclator, with for accepted species indication of distribution and, for hybrids, parents. Wastefully produced, of dubious value and moreover partly out of date (given the work of Nootemoom and others). The sectional synopsis presented is still that of Dandy (in Treseder 1978), and nowhere is there critical commentary, save on hybrids and hybridization trials.

Baranova, M. A. (1990). K voprosu o samostojatel'nosti roda *Dugandiodendron* (Magnoliaceae)/On the problem of the genus *Dugandiodendron* (Magnoliaceae) validity. Bot. Zhurn. SSSR, 75: 816-819. Ru. — Stomatographical evidence from three selected species of *Dugandiodendron* supports their inclusion in *Magnolia* or *Talauma*, in general agreement with Nooteboom (1985).

• Callaway, D. J. (1993). Magnolias. 260 pp., illus., col. pl. London: Batsford; Portland, Ore.: Timber Press. En. — A work with a North American flavour (in comparison with Treseder 1978); includes a synopsis (pp. 63-67; recent generic revisions have not been accepted, Dandy 1927 still being a basis), key to cultivated species, species descriptions (pp. 73-174), references, and chapters on breeding, hybridizers and known hybrids; three appendices and an index at end. Each chapter also has general references.

• Figlar, R. B. (1993). Stone magnolias. Arnoldia 53(2): 2-9. En. — Popular account of fossil magnolias from North America. Includes (p. 8) a series of 3 maps showing the reduction in area of sect. *Theorhodon* (to which *M. grandiflora* belongs) between 25 MYBP and the present. See also S. J. Gould, 1992. Magnolias from Moscow. Nat. Hist. 9: 10-18. [Idaho fossils remarkable for their 'ancient DNA'.]

Tobe, J. D. (1993). A molecular systematic study of eastern North American species of *Magnolia* L. vii, 144 pp. Clemson, S.C. (Unpubl. Ph.D. dissertation, Clemson University.) En. — Chapters 4 and 5 comprise a phylogenetic study based on evidence from chloroplast DNA restriction site variation; sect. *Rhytidospermum* thought to be paraphyletic as conventionally circumscribed, with a consequent exclusion of *M. acuminata* to a new section *Tulipastrum* (based on *Tulipastrum* Spach) in subgen. *Yulania* (p. 33). Chapter III comprises a systematic treatment of 6 species with keys, synonymy, typification, descriptions, exsiccatae seen, and commentary. Literature, pp. 134-144. [If ever published, drastic editing of this contribution would be in order.]

• Lozano-Contreras, G. (1994). *Dugandiodendron* y *Talauma* (Magnoliaceae) en el Neotrópico. 147 pp., illus., maps. illus., maps. Bogotá: Academic de Ciencias Exactas, Fisicas y Naturales. (Colección Jorge Alvárez Lleras, 3.) Sp. — Introduction; survey of morphology, habitats, biogeography, putative phylogeny, uses, and history of taxonomic work; well-illustrated taxonomic treatment (pp. 25-125) covering 14 *Dugandiodendron* and 31 *Talauma* species with keys, synonymy, descriptions, distribution, citation of exsiccatae, and commentary; bibliography, index and lists of taxa, exsiccatae and uses at end.

Nee, M. (1994). A new species of *Talauma* (Magnoliaceae) from Bolivia. Brittonia 46(4): 265-269, illus. En. — Description of *T. boliviana*; extensive commentary (the distinction of *Talauma* and *Dugandiodendron* is supported, 'mainly in the manner of concrescence of the carpels and their eventual dehiscence').

Nooteboom, H. P. (1994). Proposals to reject *Magnolia tomentosa* (Thymelaeaceae) and conserve *Magnolia kobus* (Magnoliaceae) with a conserved type. Taxon 43(3): 467-468. En. — A reply to Ueda (1985a); retypification of *M. kobus* DC. advocated, using an element other than a reference to *M. tomentosa* Thunb. (now generally known as *Edgeworthia tomentosa* (Thunb.) Nakai or *E. papyrifera* Sieb. & Zucc.).

• Vázquez-Garcia, J. A. (1994). *Magnolia* (Magnoliaceae) in Mexico and Central America: a synopsis. Brittonia 46(1): 1-23, illus., map. En. — Introduction (*Talauma* excluded; 4 American sections in *Magnolia* s.s. of which 3 accounted for here with one new); synopsis of 12 species (11 in sect. *Theorhodon*) and additional infraspecific taxa with keys, typification, descriptions of novelties (with citations of exsiccatae), synonymy, indication of distribution, and commentary; lists of species and exsiccatae seen at end. [Based on *idem*, 1990. Taxonomy of the genus *Magnolia* (Magnoliaceae) in Mexico and Central America. Madison, Wis. (Unpubl. M.S. thesis, University of Wisconsin, Madison). The best available modern treatment of sect. *Theorhodon*.]

Yu, Z. (1994). *Sinomanglietia*, a new genus of Magnoliaceae from China. Acta Agric. Univ. Jiangxiensis 16(2): 202-204. Ch. — Protologue and description of *S. glauca*, also new.

Pardascher, G. (1995). Magnolien. Stuttgart: Ulmer. Ge. — A handbook for enthusiasts.

Qiu, Y.-L., C. R. Parks & M. W. Chase (1995). Molecular divergence in the eastern Asia-eastern North America disjunct section *Rytidospermum* of *Magnolia* (Magnoliaceae).

Amer. J. Bot. 82: 1589-1598, illus. En. — Evidence from allozyme electrophoresis, cpDNA restriction site analysis and rbcL gene sequencing used in assessment of likely relationships among, and times of divergence of, selected species or lines thereof. The Asian *MM. hypoleuca* and *officinalis* var. *biloba* [=*M. officinalis*] were found to be rather more closely related to the American *M. tripetala* than to the two other species studied from that continent, *M. fraseri* var. *fraseri* and *M. macrophylla* var. *macrophylla*. The molecular data as well as geological and palaeoclimatic evidence suggested that separation of the Asian and American lines took place anywhere from the late Miocene to the early Pliocene. One Wagner tree is presented but there is no formal systematic treatment.

Qiu, Y.-L., M. W. Chase & C. R. Parks (1995). A chloroplast DNA phylogenetic study of the eastern Asia-eastern North America disjunct section *Rytidospermum* of *Magnolia* (Magnoliaceae). Amer. J. Bot., 82: 1582-1588, illus. En. — Chloroplast DNA sequences were sampled in all 6 species (and 4 additional infraspecific taxa) usually credited to the section (one customarily based largely on a 'whorled' leaf arrangement). Phylogenetic analysis of molecular and other evidence suggested that the section was polyphyletic, with the three Asian species closely related to only one of those in North America; moreover the group was embedded within a range of species representing the whole of Magnolioideae. The worth of characters used in the past was questioned. A Wagner tree is essayed but no formal systematic treatment presented. [See also Qiu et al., 1995. Worthy also of note is that those species shown in this paper to be closely related also mutually freely hybridise.]

International Dendrology Society (1996). Magnolias and their allies: an international symposium. 16 pp. N.p. — Programme and abstracts of a joint meeting of the International Dendrology Society and the Magnolia Society at Royal Holloway College, Egham, Surrey, United Kingdom on 12-13 April 1996.

Magnolia L., Sp. Pl.: 535 (1753).
E. Canada to Brazil, Caribbean, Nepal to New Guinea. 00 36 38 40 41 42 43 72 74 75 77 78 79 80 81 82 83 84. Nanophan. or phan.
Lassonia Buc'hoz, Pl. Nouv. Découv.: 19 (1779).
Talauma A.Juss., Gen. Pl.: 281 (1789).
Kobus Kaempf. ex Salisb., Parad. Lond.: 87 (1807).
Gwillimia Rottler ex DC., Syst. Nat. 1: 455 (1818).
Blumia Nees ex Blume, Verh. Batav. Genootsch. Kunsten 9: 147 (1823).
Aromadendron Blume, Bijdr.: 10 (1825).
Guillimia Rchb., Consp. Regn. Veg.: 193 (1828).
Sphenocarpus Wall., Numer. List: 236 (1829), nom. nud.
Lirianthe Spach, Hist. Nat. Vég. 7: 485 (1839).
Tulipastrum Spach, Hist. Nat. Vég. 7: 481 (1839).
Yulania Spach, Hist. Nat. Vég. 7: 462 (1839).
Buergeria Siebold & Zucc., Abh. Math.-Phys. Cl. Königl. Bayer. Akad. Wiss. 4: 186 (1845).
Santanderia Céspedes ex Triana & Planch., Ann. Sc. Nat. (Paris), IV, 17: 23 (1862).
Alcimandra Dandy, Bull. Misc. Inform. Kew 1927: 260 (1927).
Svenhedinia Urb., Repert. Spec. Nov. Regni Veg. 24: 3 (1927).
Parakmeria Hu & W.C.Cheng, Acta Phytotax. Sin. 1: 1 (1951).
Micheliopsis H.Keng, Quart. J. Taiwan Mus. 8: 209 (1955).
Dugandiodendron Lozano, Caldasia 11(53): 33 (1975).

Magnolia acuminata (L.) L., Syst. Nat. ed. 10: 1082 (1759).
Ontario, E. U.S.A. to Oklahoma. 72 ONT 74 OKL 75 INI NWJ OHI PEN WVA 78 ALL. Phan.
* *Magnolia virginiana* var. *acuminata* L., Sp. Pl.: 536 (1753). *Tulipastrum acuminatum* (L.) Small, Fl. S.E. U.S.: 451 (1903). *Kobus acuminata* (L.) Nieuwl., Amer. Midl. Nat. 3: 297 (1914).

var. **acuminata**

    Ontario, E. U.S.A. to Oklahoma. 72 ONT 74 OKL 75 INI NWJ OHI PEN WVA 78 ALL. Phan.

    Magnolia pensylvanica DC., Syst. Nat. 1: 453 (1817).

    Magnolia rustica DC., Syst. Nat. 1: 453 (1817).

    Magnolia decandollei Savi, Bibliot. Ital. Giorn. Lett. 16: 224 (1819). *Magnolia acuminata* var. *decandollei* (Savi) DC., Prodr. 1: 80 (1824).

    Magnolia candollei Link, Handbuch 2: 375 (1829).

    Tulipastrum americanum Spach, Hist. Nat. Vég. 7: 483 (1839).

    Tulipastrum americanum var. *vulgare* Spach, Hist. Vég. 7: 483 (1839).

    Tulipastrum acuminatum var. *aureum* Ashe, Bull. Charleston Mus. 13: 28 (1917). *Magnolia acuminata* var. *aurea* (Ashe) Ashe, Torreya 31: 38 (1931). *Magnolia acuminata* f. *aurea* (Ashe) Hardin, J.E. Mitchell Sci. Soc. 70: 306 (1954).

    Magnolia acuminata var. *ludoviciana* Sarg., Bot. Gaz. 87: 232 (1919). *Tulipastrum acuminatum* var. *ludovicianum* (Sarg.) Ashe, Bull. Torrey Bot. Club 55: 464 (1928).

var. **ozarkensis** Ashe, J.E. Mitchell Sci. Soc. 41: 269 (1926).

    NW. Arkansas, SW. Missouri, Oklahoma. 74 OKL 78 ARK MSO. Phan.

    Magnolia acuminata subsp. *ozarkensis* (Ashe) E.Murray, Kalmia 13: 9 (1983).

var. **subcordata** (Spach) Dandy, Am. J. Bot. 51: 1056 (1964).

    North Carolina, Georgia, Alabama. 78 ALA GEO NCA. Nanophan. or phan.

    Magnolia cordata Michx., Fl. Bor.-Amer. 1: 328 (1803). *Tulipastrum cordatum* (Michx.) Small, Fl. S.E. U.S.: 451 (1903). *Magnolia acuminata* var. *cordata* (Michx.) Sarg., Bot. Gaz. 87: 232 (1919). *Magnolia acuminata* subsp. *cordata* (Michx.) E.Murray, Kalmia 12: 26 (1982).

    * *Tulipastrum americanum* var. *subcordatum* Spach, Hist. Vég. 7: 483 (1839).

    Magnolia acuminata var. *alabamensis* Ashe, Torreya 31: 37 (1931).

Magnolia albosericea Chun & C.H.Tsoong, Acta Phytotax. Sin. 9: 117 (1964).

    Vietnam, China (Baoting, Lingshui, Hainan). 36 CHH CHS 41 VIE. Phan.

    Magnolia champacifolia Dandy ex Gagnep. in P.H.Lecomte, Fl. Indo-Chine, Suppl. 1: 39 (1938), nom. nud.

Magnolia allenii Standl., Publ. Field Mus. Nat. Hist., Bot. Ser. 22: 331 (1940).

    Panama (Cocle). 80 PAN. Phan.

    Talauma allenii (Standley) Lozano, Dugandiodendron Talauma Neotróp.: 78 (1994).

Magnolia amazonica (Ducke) Govaerts in D.G.Frodin & R.Govaerts, World Checklist Bibliogr. Magnoliaceae: 70 (1996).

    Brazil (Pará, Amazonas), Peru (Junin), Bolivia. 83 BOL PER 84 BZN. Phan.

    * *Talauma amazonica* Ducke, Arch. Jard. Bot. Rio de Janeiro 4: 11 (1925).

Magnolia amoena W.C.Cheng, Contrib. Biol. Lab. Chin. Assoc. Advancem. Sci., Sect. Bot. 9: 280 (1934).

    SE. China (S. Anhui, S. Jiangsu). 36 CHS. Phan.

Magnolia annamensis Dandy, J. Bot. 68: 209 (1930).

    Vietnam. 41 VIE. Phan.

    Magnolia annamensis var. *affinis* Gagnep. in ?

Magnolia arcabucoana (Lozano) Govaerts in D.G.Frodin & R.Govaerts, World Checklist Bibliogr. Magnoliaceae: 70 (1996).

    Colombia (Boyaca, Cundinamarca, Santander). 83 CLM. Phan.

    * *Talauma arcabucoana* Lozano, Fl. Colombia 1: 58 (1983).

Magnolia argyrothricha (Lozano) Govaerts in D.G.Frodin & R.Govaerts, World Checklist
    Bibliogr. Magnoliaceae: 70 (1996).
    Colombia (Boyaca, Santander). 83 CLM. Phan.
     Dugandiodendron argyrothrichum Lozano, Caldasia 11(53): 38 (1975).

Magnolia ashtonii Dandy ex Noot., Blumea 32: 363 (1987).
    Sumatra, Borneo. 42 BOR SUM. Phan.

Magnolia bintuluensis (A.Agostini) Noot., Blumea 32: 362 (1987).
    Pen. Malaysia, Sumatra (incl. Biliton), Borneo. 42 BOR MLY SUM. Phan.
     Talauma bintuluensis A.Agostini, Atti Reale Accad. Fisiocrit. Siena, X, 1: 187 (1926).
    Aromadendron nutans Dandy, Bull. Misc. Inform. Kew 1928: 183 (1928). *Magnolia nutans*
      (Dandy) H.Keng, Gard. Bull. Singapore 31: 129 (1978).

Magnolia biondii Pamp., Nuovo Giorn. Bot. Ital. n.s. 7: 275 (1910).
    China (Henan, Hubei, Shaanxi, Gansu, Sichuan). 36 CHC CHN CHS. Phan.
    Magnolia conspicua var. *fargesii* Finet & Gagnep., Bull. Soc. Bot. France 52(4): 38 (1905).
     Magnolia denudata var. *fargesii* (Finet & Gagnep.) Pamp., Bull. Soc. Tosc. Ortic. 20:
     200 (1915). *Magnolia fargesii* (Finet & Gagnep.) W.C.Cheng, J. Bot. Soc. China 1(3):
     296 (1934).
    Magnolia aulacosperma Rehder & E.H.Wilson in C.S.Sargent, Pl. Wilson. 1: 396 (1913).
    Magnolia biondii var. *ovata* T.B.Chao & T.X.Zhang, J. Henan Agric. Coll. 4: 9 (1983).
    Magnolia biondii var. *parvialabastra* T.B.Chao & al., J. Henan Agric. Coll. 4: 7 (1983).
    Magnolia biondii var. *planities* T.B.Chao & T.Z.Qiao, J. Henan Agric. Coll. 4: 10 (1983).
    Magnolia biondii var. *purpurea* T.B.Chao & Y.C.Qiao, J. Henan Agric. Coll. 4: 10 (1983).
    Magnolia honanensis B.Y.Ding & T.B.Chao, J. Henan Agric. Coll. 4: 6 (1983).
    Magnolia biondii f. *purpurascens* Y.W.Law & Z.Y.Gao, Bull. Bot. Res., Harbin 4: 192 (1984).
    Magnolia axilliflora T.B.Chao & al., Acta Agric. Univ. Henanensis 19: 360 (1985).
    Magnolia axilliflora var. *alba* T.B.Chao & al., Acta Agric. Univ. Henanensis 19: 360 (1985).
    Magnolia axilliflora var. *multitepala* T.B.Chao & al., Acta Agric. Univ. Henanensis 19: 361
     (1985).
    Magnolia biondii var. *flava* T.B.Chao & al., Acta Agric. Univ. Henanensis 19: 362 (1985).
    Magnolia biondii var. *tatitepala* T.B.Chao & J.T.Gao, Acta Agric. Univ. Henanensis 19: 363
     (1985).
    Magnolia funiushanensis T.B.Chao & al., Acta Agric. Univ. Henanensis 19: 362 (1985).
    Magnolia funiushanensis var. *purpurea* T.B.Chao & J.T.Gao, Acta Agric. Univ. Henanensis
     19: 362 (1985).

Magnolia boliviana (M.Nee) Govaerts in D.G.Frodin & R.Govaerts, World Checklist
    Bibliogr. Magnoliaceae: (1996).
    Bolivia (Cochabamba, Santa Cruz). 83 BOL. Phan.
     Talauma boliviana M.Nee, Brittonia 46: 265 (1994).

Magnolia borneensis Noot., Blumea 32: 366 (1987).
    Borneo, Philippines (Palawan). 42 BOR PHI. Phan.

Magnolia × brooklynensis Kalmb., Newslett. Amer. Magnolia Soc. 8: 7 (1972). M. acuminata
    × M. liliiflora.
    Cult. 00 CUL. Phan.

Magnolia calimaensis (Lozano) Govaerts in D.G.Frodin & R.Govaerts, World Checklist
    Bibliogr. Magnoliaceae: 70 (1996).
    Colombia (Valle). 83 CLM. Phan.
     Dugandiodendron calimaense Lozano, Dugandiodendron Talauma Neotróp.: 35 (1994).

Magnolia calophylla (Lozano) Govaerts in D.G.Frodin & R.Govaerts, World Checklist
Bibliogr. Magnoliaceae: 70 (1996).
Colombia (Nariño). 83 CLM. Phan.

* *Dugandiodendron calophyllum* Lozano, Caldasia 12(58): 283 (1978).

Magnolia campbellii Hook.f. & Thomson, Fl. Ind. 1: 77 (1855).
E. Nepal, Sikkim, Bhutan, N. Assam, SE. Tibet, N. Burma, SW. China (W. Sichuan, N. & W.
Yunnan). 36 CHC CHT 40 ASS BHU NEP 41 BMA. Phan.
Magnolia mollicomata W.W.Sm., Notes Roy. Bot. Gard. Edinburgh 12: 211 (1920). *Magnolia
campbellii* var. *mollicomata* (W.W.Sm.) F.S.Ward, Gard. Chron. 137: 238 (1955). *Magnolia
campbellii* subsp. *mollicomata* (W.W.Sm.) Johnstone, Asiatic Magnolias Cult.: 53 (1955).
Magnolia campbellii var. *alba* Tresder, Magnolias: 90 (1987).

Magnolia cararensis (Lozano) Govaerts in D.G.Frodin & R.Govaerts, World Checklist
Bibliogr. Magnoliaceae: 70 (1996).
Colombia (Norte de Santander). 83 CLM. Phan.
* *Dugandiodendron cararense* Lozano, Dugandiodendron Talauma Neotróp.: 52 (1994).

Magnolia caricifragrans (Lozano) Govaerts in D.G.Frodin & R.Govaerts, World Checklist
Bibliogr. Magnoliaceae: 70 (1996).
Colombia (Boyaca, Cundinamacca, Norte de Santander). 83 CLM. Phan.
* *Talauma caricifragrans* Lozano, Mutisia 36: 2 (1972).

Magnolia carsonii Dandy ex Noot., Blumea 32: 348 (1987).
Borneo. 42 BOR. Phan.

var. **carsonii**
Borneo (Sabah: Mt. Kinabalu). 42 BOR. Phan.

var. **drymifolia** Noot., Blumea 32: 351 (1987).
Borneo. 42 BOR. Nanophan. or phan.

Magnolia cathcartii (Hook.f. & Thomson) Noot., Blumea 31: 88 (1985).
Sikkim, Assam, S. & SE. Tibet, N. Burma, Vietnam. 36 CHT 40 ASS BHU 41 BMA VIE. Phan.
* *Michelia cathcartii* Hook.f. & Thomson, Fl. Ind. 1: 79 (1855). *Sampacca cathcartii*
(Hook.f. & Thomson) Kuntze, Revis. Gen. Pl.: 6 (1891). *Alcimandra cathcartii* (Hook.f.
& Thomson) Dandy, Bull. Misc. Inform. Kew 1927: 260 (1927).

Magnolia cespedesii (Triana & Planch.) Govaerts in D.G.Frodin & R.Govaerts, World
Checklist Bibliogr. Magnoliaceae: 70 (1996).
Colombia (Cundinamacca, Boyaca). 83 CLM. Phan.
* *Talauma cespedesii* Triana & Planch., Ann. Sc. Nat. (Paris), IV, 17: 23 (1862).

Magnolia championii Benth., Fl. Hongk.: 8 (1861). *Magnolia pumila* var. *championii*
(Benth.) Finet & Gagnep., Bull. Soc. Bot. France 52(4): 36 (1905). *Magnolia liliifera* var.
championii (Benth.) Pamp., Bull. Soc. Tosc. Ortic. 4: 136 (1916).
Vietnam, Hainan, Taiwan, S. China (SE. Yunann, Guangdong, Hong Kong, Guangxi,
Guizhou). 36 CHC CHS 38 THA 41 VIE. Nanophan. or phan.
Talauma fistulosa Finet & Gagnep., Bull. Soc. Bot. France 52(4): 31 (1906). *Magnolia
fistulosa* (Finet & Gagnep.) Dandy, Notes Roy. Bot. Gard. Edinburgh 16: 124 (1928).
Magnolia paenetalauma Dandy, J. Bot. 68: 206 (1930).
Magnolia talaumoides Dandy, J. Bot. 68: 208 (1930).
Magnolia tenuicarpella H.T.Chang, Acta Sci. Nat. Univ. Sunyatseni 1961: 56 (1961).
Magnolia odoratissima Y.W.Law & R.Z.Zhou, Bull. Bot. Res., Harbin 6: 139 (1986).

Magnolia chimantensis Steyerm. & Maguire, Mem. New York Bot. Gard. 17(1): 443 (1967).
Dugandiodendron chimantense (Steyerm. & Maguire) Lozano, Caldasia 12(56): 9 (1977).
Venezuela (Bolivar). 82 VEN. Phan.

Magnolia chocoensis (Lozano) Govaerts in D.G.Frodin & R.Govaerts, World Checklist
Bibliogr. Magnoliaceae: 70 (1996).
Colombia (Choco, Risaralda). 83 CLM. Phan.
* *Talauma chocoensis* Lozano, Fl. Colombia 1: 67 (1983).

Magnolia clemensiorum Dandy, J. Bot. 68: 207 (1930).
Vietnam. 41 VIE. Phan.

Magnolia coco (Lour.) DC., Syst. Nat. 1: 459 (1817).
N. Vietnam, Taiwan, S. China (Guangdong, Guangxi, Guizhou, Fukien, Fujian, Zhejiang).
36 CHC CHS 38 TAI 41 VIE. Nanophan. or phan.
* *Liriodendron coco* Lour., Fl. Cochinch.: 347 (1790). *Talauma coco* (Lour.) Merr., Sp.
Blancoan.: 12 (1918).
Gwillimia indica Rottler ex DC., Syst. Nat. 1: 458 (1818).
Liriopsis pumila Spach ex Baill., Adansonia 7: 4 (1866).

Magnolia colombiana (Little) Govaerts in D.G.Frodin & R.Govaerts, World Checklist
Bibliogr. Magnoliaceae: 70 (1996).
Colombia (Huila). 83 CLM. Phan.
* *Talauma colombiana* Little, Phytologia 19: 292 (1970). *Dugandiodendron colombianum*
(Little) Lozano, Caldasia 11(53): 43 (1975).

Magnolia cristalensis Bisse, Repert. Spec. Nov. Regni Veg. 85: 588 (1974).
Cuba. 81 CUB. Phan.

Magnolia cubensis Urb., Symb. Antill. 1: 307 (1899).
Cuba. 81 CUB. Phan.

subsp. **cacuminicola** (Bisse) G.Klotz, Wiss. Z. Friedrich-Schiller-Univ. Jena, Math.-
Naturwiss. Reihe 29: 464 (1980).
Cuba. 81 CUB. Phan.
* *Magnolia cacuminicola* Bisse, Repert. Spec. Nov. Regni Veg. 85: 587 (1974).

subsp. **cubensis**
Cuba (Sierra Maestra). 81 CUB. Phan.
Magnolia cubensis subsp. *acunae* Imkhan., Novosti Sist. Vyssh. Rast. 11: 180 (1974).
Magnolia cubensis var. *baracoensis* Imkhan., Novosti Sist. Vyssh. Rast. 11: 179 (1974).

Magnolia cylindrica E.H.Wilson, J. Arnold Arbor. 8: 109 (1927).
E. China (Anhui, Fujian, Jiangxi, Zhejiang). 36 CHS. Phan.

Magnolia dawsoniana Rehder & E.H.Wilson. in C.S.Sargent, Pl. Wilson. 1: 397 (1913).
SW. China (C. & S. Sichuan, N. Yunnan). 36 CHC. Phan.

Magnolia dealbata Zucc., Abh. Math.-Phys. Cl. Königl. Bayer. Akad. Wiss. 2: 373 (1836).
Magnolia macrophylla var. *dealbata* (Zucc.) D.L.Johnson, Baileya 23: 56 (1989).
Mexico, Cuba, Hispaniola, Puerto Rico. 79 MXE MXS 81 CUB DOM HAI PUE. Phan.

Magnolia delavayi Franch., Pl. Delavay.: 33 (1889).
SW. China (Sichuan, Yunnan). 36 CHC. Phan.

Magnolia denudata Desr. in J.B.A.M.de Lamarck, Encycl. 3: 675 (1792), nom. cons. prop.
(to be published)
Magnolia purpurea var. *denudata* (Desr.) Loudon in ? *Magnolia obovata* var. *denudata* (Desr.)
DC., Syst. Nat. 1: 457 (1817).
E. & S. China (Anhui, Hunan, Fujian, Jiangsu, Zhejiang, Guangdong, Guizhou). 36 CHC
CHS. Phan.

Magnolia conspicua var. *purpurascens* Rehder & E.H.Wilson in ?, nom. inval.

Magnolia conspicua var. *rosea* Veits in ?

Lassonia heptapeta Buc'hoz, Pl. Nouv. Découv.: 19 (1779). *Magnolia heptapeta* (Buc'hoz)
    Dandy, J. Bot. 72: 103 (1934).

Magnolia precia Corrêa ex Vent., Jard. Malmaison: 24 (1803), nom. nud.

Magnolia conspicua Salisb., Parad. Lond.: 38 (1806). *Yulania conspicua* (Salisb.) Spach, Hist.
    Nat. Vég. 7: 464 (1839).

Magnolia yulan Desf., Hist. Arbr. France 2: 6 (1809). *Michelia yulan* (Desf.) Kostel., Allg.
    Med.-Pharm. Fl. 5: 1700 (1836). *Gwillimia yulan* (Desf.) Kostel., Allg. Med.-Pharm. Fl. 5:
    1700 (1836).

Magnolia alexandrina Steud., Nomencl. Bot., ed. 2, 2: 89 (1841).

Magnolia citriodora Steud., Nomencl. Bot., ed. 2, 2: 89 (1841).

Magnolia cyathiformis Rinz ex K.Koch, Dendrologie 1: 376 (1869).

Magnolia spectabilis G.Nicholson, Hand-List of Trees and Shrubs 1: 15 (1894), nom. inval.

Magnolia superba G.Nicholson, Hand-List of Trees and Shrubs 1: 15 (1894), nom. inval.

Magnolia triumphans G.Nicholson, Hand-List of Trees and Shrubs 1: 15 (1894), nom. inval.

Magnolia denudata var. *pyramidalis* T.B.Chao & Z.X.Chen, J. Henan Agric. Coll. 4: 11
    (1983).

Magnolia denudata var. *angustitepala* T.B.Chao & Z.S.Chun, Acta Agric. Univ. Henanensis
    19: 363 (1985).

Magnolia dixonii (Little) Govaerts in D.G.Frodin & R.Govaerts, World Checklist Bibliogr.
    Magnoliaceae: 70 (1996).
    Ecuador (Esmeraldas). 83 ECU. Phan.
     * *Talauma dixonii* Little, Phytologia 18: 457 (1969).

Magnolia dodecapetala (Lam.) Govaerts in D.G.Frodin & R.Govaerts, World Checklist
    Bibliogr. Magnoliaceae: 70 (1996).
    Martinique, Trinidad, Guadeloupe, St. Vincent, Dominica. 80 LEE TRT WIN. Phan.
     * *Annona dodecapetala* Lam., Encycl. 2: 127 (1786). *Talauma dodecapetala* (Lam.) Urb.,
       Repert. Spec. Nov. Regni Veg. 15: 306 (1918).
     Magnolia plumieri Sw., Prodr.: 87 (1797). *Talauma plumieri* (Sw.) DC., Prodr. 1: 81 (1824).
     Talauma caerulea J.St.-Hil., Expos. Fam. Nat. 2: 76 (1805).
     Magnolia linguifolia L. ex Descourt., Fl. Méd. Antilles 2: 140 (1822).
     Magnolia fatiscens Rich. ex DC., Prodr. 1: 82 (1824).
     Talauma coerulea Steud., Nomencl. Bot., ed. 2, 2: 660 (1841).

Magnolia domingensis Urb., Repert. Spec. Nov. Regni Veg. 13: 447 (1914).
    Haiti. 81 HAI. Nanophan. or phan

Magnolia ekmanii Urb., Ark. Bot. 23A(11): 12 (1931).
    Haiti. 81 HAI. Phan.

Magnolia elegans (Blume) H.Keng, Gard. Bull. Singapore 31: 129 (1978).
    Pen. Malaysia, Sumatra (incl. Bangka), W. Java. 42 JAW MLY SUM. Phan.
     * *Aromadendron elegans* Blume, Bijdr.: 10 (1825). *Talauma elegans* (Blume) Miq., Ann.
       Mus. Bot. Lugduno-Batavi 4: 70 (1868).
     Aromadendron glaucum Korth., Ned. Kruidk. Arch. 2(2): 98 (1851). *Talauma glaucum*
       (Korth.) Miq., Ann. Mus. Bot. Lugduno-Batavi 4: 70 (1868). *Magnolia glauca* (Korth.)
       Pierre, Fl. Forest. Cochinch.: 2 (1880). *Talauma elegans* var. *glauca* (Korth.) P.Parm.,
       Bull. Sc. France Belgique 27: 277, 336 (1896). *Aromadendron elegans* var. *glauca*
       (Korth.) Dandy, Bull. Misc. Inform. Kew 1928: 183 (1928).
     Manglietia oortii Korth., Ned. Kruidk. Arch. 2(2): 97 (1851).

Magnolia emarginata Urb. & Ekman, Ark. Bot. 23A(11): 11 (1931).
    N. Haiti. 81 HAI. Phan.

Magnolia espinalii (Lozano) Govaerts in D.G.Frodin & R.Govaerts, World Checklist
Bibliogr. Magnoliaceae: 70 (1996).
Colombia (Antioquia). 83 CLM. Phan.
  * *Talauma espinalii* Lozano, Fl. Colombia 1: 70 (1983).

Magnolia fraseri Walter, Fl. Carol.: 159 (1788).
E. U.S.A. to Texas. 75 WVA 77 TEX 78 ALA FLA GEO KTY LOU MSI NCA SCA TEN VRG.
Phan.

  var. **fraseri**
E. & SE. U.S.A. 75 WVA 78 GEO KTY NCA SCA TEN VRG. Phan.
Magnolia auriculata Desr. in J.B.A.M.de Lamarck, Encycl. 3: 673 (1792).
Magnolia auricularis Salisb., Parad. Lond.: 43 (1806).

  var. **pyramidata** (Bartram) Pamp., Bull. Soc. Tosc. Ortic. 40: 230 (1915).
SE. U.S.A. to Texas. 77 TEX 78 ALA FLA GEO LOU MSI SCA. Phan.
  * *Magnolia pyramidata* Bartram, Travels Carolina: 408 (1791). *Magnolia auriculata* var.
    pyramidata (Bartram) Nutt., Gen. N. Americ. Pl. 2: 18 (1818). *Magnolia fraseri* subsp.
    pyramidata (Bartram) E.Murray, Kalmia 13: 9 (1983).

Magnolia georgii (Lozano) Govaerts in D.G.Frodin & R.Govaerts, World Checklist Bibliogr.
Magnoliaceae: 70 (1996).
Colombia (Boyaca, Santander). 83 CLM. Phan.
  * *Talauma georgii* Lozano, Fl. Colombia 1: 76 (1983).

Magnolia gigantifolia (Miq.) Noot., Blumea 32: 377 (1987).
Sumatra (incl. Bangka), Borneo. 42 BOR SUR. Phan.
  * *Talauma gigantifolia* Miq., Fl. Ned. Ind. 1(2): 15 (1858).
  Talauma megalophylla Merr., J. Straits Branch Roy. Asiat. Soc. 85: 172 (1922).
  Talauma magna A.Agostini, Atti Reale Accad. Fisiocrit. Siena, X, 1: 192 (1926).

Magnolia gilbertoi (Lozano) Govaerts in D.G.Frodin & R.Govaerts, World Checklist
Bibliogr. Magnoliaceae: 70 (1996).
Colombia (Risarolde, Valle). 83 CLM. Phan.
  * *Talauma gilbertoi* Lozano, Fl. Colombia 1: 73 (1983).

Magnolia globosa Hook.f. & Thomson, Fl. Ind. 1: 77 (1855). *Yulania japonica* var. *globosa*
(Hook.f. & Thomson) P.Parm., Bull. Sc. France Belgique 27: 258 (1895).
Nepal, Bhutan, NE. Assam, SE. Tibet, N. Burma, SW. China (SW. Sichuan, NW. Yunnan).
36 CHC CHT 40 ASS BHU NEP 41 BMA. Nanophan. or phan.
Magnolia tsarongensis W.W.Sm. & Forrest, Notes Roy. Bot. Gard. Edinburgh 12: 215 (1920).

Magnolia gloriensis (Pittier) Govaerts in D.G.Frodin & R.Govaerts, World Checklist
Bibliogr. Magnoliaceae: 71 (1996).
Costa Rica, Nicaragua, Panama. 80 COS NIC PAN. Phan.
  * *Talauma gloriensis* Pittier, Contrib. U.S. Natl. Herb. 13: 94 (1910).

Magnolia grandiflora L., Syst. Nat. ed. 10: 1082 (1759).
SE. U.S.A. to Texas. 77 TEX 78 ALA FLA GEO LOU MSI NCA SCA. Phan.
Magnolia microphylla Ser. in ?
Magnolia obtusifolia .
Magnolia tardiflora Ser. in ?
Magnolia tomentosa Ser. in ?
Magnolia virginiana var. *foetida* L., Sp. Pl.: 536 (1753). *Magnolia foetida* (L.) Sarg., Gard. &
  Forest 2: 615 (1889).
Magnolia grandiflora var. *lanceolata* Aiton, Hortus Kew. 2: 251 (1789). *Magnolia*
  grandiflora f. *lanceolata* (Aiton) Rehder, Bibl. Cult. Trees: 180 (1949).
Magnolia grandiflora var. *elliptica* W.T.Aiton, Hortus Kew. 3: 329 (1811).

Magnolia grandiflora var. *obovata* W.T.Aiton, Hortus Kew. 3: 329 (1811).

Magnolia longifolia Sweet, Hort. Brit. ed. 1: 11 (1826).

Magnolia elliptica Link., Handbuch 2: 375 (1829).

Magnolia lanceolata Link, Handbuch 2: 375 (1829).

Magnolia obovata Aiton ex Link, Handbuch 2: 375 (1829).

Magnolia maxima Lodd. ex G.Don in J.C.Loudon, Hort. Brit.: 226 (1830).

Magnolia grandiflora var. *exoniensis* Loud., Arbor. Frutic. Brit. 1: 261 (1839).

Magnolia lacunosa Raf., Autik. Bot.: 78 (1840).

Magnolia grandiflora f. *galissoniensis* K.Koch, Dendrologie 1: 368 (1869).

Magnolia ferruginea W.Watson, Bull. Misc. Inform. Kew 1889: 305 (1889).

Magnolia hartwegii G.Nicholson, Hand-List of Trees and Shrubs 1: 17 (1894), nom. inval.

Magnolia hartwicus G.Nicholson, Hand-List of Trees and Shrubs 1: 17 (1894), nom. inval.

Magnolia stricta G.Nicholson, Hand-List of Trees and Shrubs 1: 17 (1894), nom. inval.

Magnolia angustifolia Millais, Magnolias: 55, 83 (1927).

Magnolia exoniensis Millais, Magnolias: 59 (1927).

Magnolia galissoniensis Millais, Magnolias: 60 (1927).

Magnolia gloriosa Millais, Magnolias: 61 (1927).

Magnolia praecox Millais, Magnolias: 69 (1927).

Magnolia pravertiana Millais, Magnolias: 69 (1927).

Magnolia rotundifolia Millais, Magnolias: 70 (1927).

Magnolia foetida f. *margaretta* Ashe, Torreya 31: 37 (1931).

Magnolia foetida f. *parvifolia* Ashe, Torreya 31: 37 (1931).

Magnolia griffithii Hook.f. & Thomson in J.D.Hooker, Fl. Brit. Ind. 1: 41 (1872). *Michelia griffithii* (Hook.f. & Thomson) Finet & Gagnep., Bull. Soc. Bot. France 52(4): 42 (1906). Assam, N. Burma. 40 ASS 41 BMA. Phan.

Magnolia guatapensis (Lozano) Govaerts in D.G.Frodin & R.Govaerts, World Checklist Bibliogr. Magnoliaceae: 71 (1996).
Colombia (Antioquia). 83 CLM. Phan.
  * *Dugandiodendron guatapense* Lozano, Dugandiodendron Talauma Neotróp.: 50 (1994).

Magnolia guatemalensis Donn.Sm., Bot. Gaz. 47: 253 (1909).
C. America. 80 ELS HON GUA. Phan.

subsp. **guatemalensis**
  Guatemala. 80 GUA. Phan.

subsp. **hondurensis** (Molina) Vázquez, Brittonia 46: 6 (1994).
  El Salvador, Honduras. 80 ELS HON. Phan.
  * *Magnolia hondurensis* A.M.Molina, Ceiba 18: 95 (1974).

Magnolia gustavii King, Ann. Roy. Bot. Gard. (Calcutta) 3(2): 209 (1891).
Assam (Makum Forest). 40 ASS. Phan.
  Michelia gustavii King, Ann. Roy. Bot. Gard. (Calcutta) 3(2): 209 (1891).

Magnolia hamorii Howard, Bull. Torrey Bot. Club 75: 351 (1948).
Dominican Rep. (Barahona). 81 DOM. Phan.

Magnolia henaoi (Lozano) Govaerts in D.G.Frodin & R.Govaerts, World Checklist Bibliogr. Magnoliaceae: 71 (1996).
Colombia (Huila). 83 CLM. Phan.
  * *Talauma henaoi* Lozano, Fl. Colombia 1: 78 (1983).

Magnolia henryi Dunn, J. Linn. Soc., Bot. 35: 484 (1903).
N. Burma, Thailand, Laos, SW. China (SW. Yunnan). 36 CHC 41 BMA LAO THA. Phan.
Talauma kerrii Craib, Bull. Misc. Inform. Kew 1922: 226 (1922).
Manglietia wangii Hu, Bull. Fan Mem. Inst. Biol. 8: 33 (1937).

Magnolia hernandezii (Lozano) Govaerts in D.G.Frodin & R.Govaerts, World Checklist Bibliogr. Magnoliaceae: 71 (1996).
Colombia (Antioquia, Quindio, Valle). 83 CLM. Phan.
   * *Talauma hernandezii* Lozano, Mutisia 37: 11 (1972).

Magnolia iltisiana Vázquez, Brittonia 46: 7 (1994).
Mexico (Jalisco, Michoacan, Guerrero). 79 MXS. Phan.

Magnolia irwiniana (Lozano) Govaerts in D.G.Frodin & R.Govaerts, World Checklist Bibliogr. Magnoliaceae: 71 (1996).
Brazil (Goias). 84 BZC. Phan.
   * *Talauma irwiniana* Lozano, Revista Acad. Colomb. Ci. Exact. 66: 580 (1990).

Magnolia kachirachirai (Kaneh. & Yamam.) Dandy, Bull. Misc. Inform. Kew 1927: 264 (1927).
SE. Taiwan. 38 TAI. Phan.
   * *Michelia kachirachirai* Kaneh. & Yamam. in B.Hayata, Icon. Pl. Formos., Suppl. 2: 14 (1926). *Micheliopsis kachirachirai* (Kaneh. & Yamam.) H.Keng, Quart. J. Taiwan Mus. 8: 210 (1955). *Parakmeria kachirachirai* (Kaneh. & Yamam.) Y.W.Law in W.C.Cheng, Sylva Sinica 1: 473 (1983). *Parakmeria kachirachirai* (Kaneh. & Yamam.) Y.W.Law, Acta Phytotax Sin. 34: 91 (1996).

Magnolia katiorum (Lozano) Govaerts in D.G.Frodin & R.Govaerts, World Checklist Bibliogr. Magnoliaceae: 71 (1996).
Colombia (Antioquia). 83 CLM. Phan.
   * *Talauma katiorum* Lozano, Fl. Colombia 1: 84 (1983).

Magnolia kobus DC., Syst. Nat. 1: 456 (1817). *Yulania kobus* (DC.) Spach, Hist. Nat. Vég. 7: 467 (1839).
Japan, S. Korea (incl. Cheju Do). 38 JAP KOR. Nanophan. or phan.
   Michelia gracilis Kostel., Allg. Med.-Pharm. Fl. 5: 1701 (1836).
   Buergeria obovata Siebold & Zucc., Abh. Math.-Phys. Cl. Königl. Bayer. Akad. Wiss. 4: 187 (1845). *Talauma obovata* (Siebold & Zucc.) Benth. & Hook.f. ex Hance, J. Bot. 20: 2 (1882).
   Magnolia thurberi G.Nicholson, Hand-List of Trees and Shrubs 1: 17 (1894), nom. inval.
   Magnolia kobus var. *borealis* Sarg., Trees & Shrubs 2: 57 (1908). *Magnolia borealis* (Sarg.) Kudô, Medic. Pl. Hokkaido: 47 (1922).
   Magnolia praecocossima Koidz., Bot. Mag. (Tokyo) 43: 386 (1929), nom. nud.
   Magnolia pseudokobus S.Abe & Akasawa, Bull. Kochi Women's Coll. 2: 104, 110 (1954). A triploid of M. kobus.

Magnolia lasia Noot., Blumea 32: 377 (1987).
Borneo (Sarawak, Sabah, E. Kalimantan). 42 BOR. Phan.

Magnolia lenticellatum (Lozano) Govaerts in D.G.Frodin & R.Govaerts, World Checklist Bibliogr. Magnoliaceae: 71 (1996).
Colombia (Antioquia). 83 CLM. Phan.
   * *Dugandiodendron lenticellata* Lozano, Dugandiodendron Talauma Neotróp.: 46 (1994).

Magnolia liliifera (L.) Baill., Hist. Pl. 1: 141 (1868).
China, Trop. Asia. 36 CHC CHH CHT 40 ASS BHU NEP 41 BMA CBD THA VIE 42 BOR JAW LSI MOL NWG PHI SUL SUM 43 AND. Nanophan. or phan.
   * *Liriodendron liliiferum* L., Sp. Pl. ed. 2: 755 (1762). *Talauma liliifera* (L.) Kuntze, Revis. Gen. Pl.: 6 (1891).

var. **angatensis** (Blanco) Govaerts in D.G.Frodin & R.Govaerts, World Checklist Bibliogr.
Magnoliaceae: 71 (1996).
Philippines, Moluccas (Talaud). 42 MOL PHI. Phan.
* *Magnolia angatensis* Blanco, Fl. Filip.: 859 (1837). *Talauma angatensis* (Blanco) Fern.-Vill.
in F.M.Blanco, Fl. Philipp., ed. 3 Nov. App.: 3 (1880). *Magnolia candollei* var.
angatensis (Blanco) Noot., Blumea 32: 375 (1987).
Talauma mutabilis Fern.-Vill. in F.M.Blanco, Fl. Philipp., ed. 3 Nov. App.: 3 (1880).
Talauma villariana Rolfe, J. Linn. Soc., Bot. 21: 307 (1884). *Magnolia villariana* (Rolfe)
D.C.S.Raju & M.P.Nayar, Indian J. Bot. 3: 171 (1980).
Talauma luzonensis Warb. in J.R.Perkins, Fragm. Fl. Philipp.: 171 (1904).
Talauma grandiflora Merr., Publ. Bur. Sci. Gov. Lab. 29: 13 (1905).
Talauma oblongata Merr., Publ. Bur. Sci. Gov. Lab. 35: 8 (1906).

var. **beccarii** (Ridl.) Govaerts in D.G.Frodin & R.Govaerts, World Checklist Bibliogr.
Magnoliaceae: 71 (1996).
Borneo. 42 BOR. Phan.
* *Talauma beccarii* Ridl., Bull. Misc. Inform. Kew 1912: 381 (1912). *Magnolia candollei* var.
beccarii (Ridl.) Noot., Blumea 32: 375 (1987).

var. **liliifera**
Sikkim, Assam, Thailand, Cambodia, Vietnam, Hainan, Sumatra to New Guinea,
Andaman Is. 36 CHH 40 ASS BHU 41 CBD THA VIE 42 BOR JAW LSI MOL NWG PHI
SUL SUM 43 AND. Nanophan.
Magnolia pumila Andrews, Bot. Repos.: 226 (1802). *Talauma pumila* (Andrews) Blume,
Fl. Javae 19-20: 38 (1829).
Blumia candollei (Blume) Nees, Verh. Batav. Genootsch. Kunsten 9: 147 (1823).
Talauma candollei Blume, Verh. Batav. Genootsch. Kunsten 9: 147 (1823). *Blumia*
candollei (Blume) Nees, Verh. Batav. Genootsch. Kunsten 9: 147 (1823). *Magnolia*
candollei (Blume) H.Keng, Gard. Bull. Singapore 31: 129 (1978).
Talauma candollei var. *latifolia* Blume, Bijdr.: 9 (1825).
Talauma rumphii Blume, Bijdr.: 10 (1825). *Magnolia rumphii* (Blume) Spreng., Syst. Veg.
4(2): 217 (1827).
Magnolia odoratissima Reinw. ex Blume, Fl. Javae 19-20: 32 (1829).
Talauma mutabilis Blume, Fl. Javae 19-20: 35 (1829). *Magnolia mutabilis* (Blume)
H.J.Chowdhery & P.Daniel, Indian J. Forest. 4: 64 (1981).
Talauma mutabilis var. *acuminata* Blume, Fl. Javae 19-20: 37 (1829).
Talauma mutabilis var. *longifolia* Blume, Fl. Javae 19-20: 36 (1829). *Talauma longifolia*
(Blume) Ridl., J. Fed. Malay States Mus. 7: 38 (1916).
Talauma mutabilis var. *splendens* Blume, Fl. Javae 19-20: 38 (1829).
Magnolia splendens Reinw. ex Blume, Fl. Javae 19-20: 38 (1825), pro syn.
Talauma rabaniana Hook.f. & Thomson, Fl. Ind. 1: 75 (1855). *Magnolia rabaniana*
(Hook.f. & Thomson) D.C.S.Raju & M.P.Nayar, Indian J. Bot. 3: 171 (1980).
Talauma rubra Miq., Fl. Ned. Ind. 1(2): 14 (1858).
Manglietia celebica Miq., Ann. Mus. Bot. Lugduno-Batavi 4: 72 (1868). *Talauma*
miqueliana Dandy, Bull. Misc. Inform. Kew 1927: 262 (1927).
Talauma macrophylla Blume ex Miq., Ann. Mus. Bot. Lugduno-Batavi 4: 68 (1868).
Manglietia sebassa King, J. Asiat. Soc. Bengal 58(2): 370 (1889). *Talauma sebassa* (King)
Miq. ex Dandy, Bull. Misc. Inform. Kew 1928: 192 (1928).
Talauma andamanica King, J. Asiat. Soc. Bengal 58(2): 372 (1889). *Magnolia andamanica*
(King) D.C.S.Raju & M.P.Nayar, Indian J. Bot. 3: 171 (1980).
Talauma forbesii King, J. Asiat. Soc. Bengal 58(2): 373 (1889).
Talauma kunstleri King, J. Asiat. Soc. Bengal 58(2): 373 (1889).
Magnolia forbesii King, Ann. Roy. Bot. Gard. (Calcutta) 3(2): 206 (1891), pro syn.
Magnolia kunstleri King, Ann. Roy. Bot. Gard. (Calcutta) 3(2): 204 (1891), pro syn.
Talauma inflata P.Parm., Bull. Sc. France Belgique 27: 208, 273 (1896).
Talauma javanica P.Parm., Bull. Sc. France Belgique 27: 208, 274 (1896).
Talauma gitingensis Elmer, Leafl. Philipp. 4: 1479 (1912).

Talauma oreadum Diels, Bot. Jahrb. Syst. 54: 240 (1916). *Aromadendron oreadum* (Diels) Kaneh. & Hatus., Bot. Mag. (Tokyo) 57: 147 (1943).

Talauma reticulata Merr., Philipp. J. Sci. 17: 249 (1920 publ. 1921).

Talauma borneensis Merr., J. Straits Branch Roy. Asiat. Soc. 85: 173 (1922).

Talauma sumatrana A.Agostini, Atti Reale Accad. Fisiocrit. Siena, X, 1: 189 (1926).

Talauma undulatifolia A.Agostini, Atti Reale Accad. Fisiocrit. Siena, X, 1: 188 (1926).

Magnolia pachyphylla Dandy, Bull. Misc. Inform. Kew 1928: 186 (1928).

Talauma athliantha Dandy, Bull. Misc. Inform. Kew 1928: 189 (1928).

Talauma gitingensis var. *glabra* Dandy, Bull. Misc. Inform. Kew 1928: 189 (1928).

Talauma gitingensis var. *rotundata* Dandy, Bull. Misc. Inform. Kew 1928: 190 (1928).

Talauma gracilior Dandy, Bull. Misc. Inform. Kew 1928: 190 (1928).

Talauma peninsularis Dandy, Bull. Misc. Inform. Kew 1928: 192 (1928).

Talauma soembensis Dandy, Bull. Misc. Inform. Kew 1928: 193 (1928).

Magnolia craibiana Dandy, Bull. Misc. Inform. Kew 1929: 105 (1929).

Talauma siamensis Dandy, Bull. Misc. Inform. Kew 1929: 105 (1929). *Magnolia siamensis* (Dandy) H.Keng, Gard. Bull. Singapore 31: 129 (1978).

Magnolia thamnodes Dandy, J. Bot. 68: 208 (1930). *Manglietia thamnodes* (Dandy) Gagnep. in P.H.Lecomte, Fl. Indo-Chine, Suppl. 1: 35 (1939). *Talauma thamnodes* (Dandy) Tiep, Repert. Spec. Nov. Regni Veg. 91: 507 (1980).

Talauma nhatrangensis Dandy, J. Bot. 68: 210 (1930).

Magnolia eriostepta var. *poilanei* Dandy ex Humbert in P.H.Lecomte, Fl. Indo-Chine, Suppl. 1: 40 (1938).

var. **obovata** (Korth.) Govaerts in D.G.Frodin & R.Govaerts, World Checklist Bibliogr. Magnoliaceae: 71 (1996).

    Nepal, Sikkim, Tibet, Assam, Bhutan, Thailand, Pen. Malaysia, SW. China, Borneo. 36 CHC CHT 40 ASS BHU NEP 41 BMA THA 42 BOR MLY. Phan.

   * *Talauma obovata* Korth., Ned. Kruidk. Arch. 2(2): 89 (1851). *Magnolia candollei* var. *obovata* (Korth.) Noot., Blumea 32: 374 (1987).

Talauma hodgsonii Hook.f. & Thomson, Fl. Ind. 1: 74 (1855). *Magnolia hodgsonii* (Hook.f. & Thomson) H.Keng, Gard. Bull. Singapore 31: 129 (1978).

Talauma betongensis Craib, Bull. Misc. Inform. Kew 1925: 7 (1925). *Magnolia betongensis* (Craib) H.Keng, Gard. Bull. Singapore 31: 129 (1978).

Talauma oblanceolata Ridl., Fl. Malay. Penins. 5: 286 (1925).

Talauma levissima Dandy, Bull. Misc. Inform. Kew 1928: 191 (1928).

Talauma sclerophylla Dandy, J. Bot. 66: 47 (1928).

var. **singapurensis** (Ridl.) Govaerts in D.G.Frodin & R.Govaerts, World Checklist Bibliogr. Magnoliaceae: 71 (1996).

    Pen. Malaysia (incl. Singapore), Sumatra (incl. Bangka), Borneo. 42 BOR MLY SUM. Phan.

   * *Talauma singapurensis* Ridl., Bull. Misc. Inform. Kew 1914: 323 (1914). *Magnolia singapurensis* (Ridl.) H.Keng, Gard. Bull. Singapore 31: 129 (1978). *Magnolia candollei* var. *singapurensis* (Ridl.) Noot., Blumea 32: 376 (1987).

Talauma kuteinensis A.Agostini, Atti Reale Accad. Fisiocrit. Siena, X, 1: 191 (1926).

Magnolia liliiflora Desr. in J.B.A.M.de Lamarck, Encycl. 3: 675 (1792), nom. cons. prop. (to be published)

    C. & SW. China (Hubei, Yunnan). 36 CHC. Nanophan.

   Lassonia quinquepeta Buc'hoz, Pl. Nouv. Découv.: 19 (1779). *Magnolia quinquepeta* (Buc'hoz) Dandy, J. Bot. 72: 103 (1934).

Magnolia purpurea Curtis, Bot. Mag.: 390 (1797). *Yulania japonica* var. *purpurea* (Curtis) P.Parm., Bull. Sc. France Belgique 27: 258 (1896).

Magnolia discolor Vent., Jard. Malmaison: 24 (1803).

Magnolia gracilis Salisb., Parad. Lond.: 87 (1807). *Magnolia liliiflora* var. *gracilis* (Salisb.) Rehder in L.H.Bailey, Stand. Cycl. Hort. 4: 1968 (1916).

Magnolia atropurpurea Steud., Nomencl. Bot.: 504 (1824).

Yulania japonica Spach, Hist. Nat. Vég. 7: 466 (1839).

Talauma sieboldii Miq., Ann. Mus. Bot. Lugduno-Batavi 2: 257 (1866).

Magnolia × *soulangeana* var. *nigra* G.Nicholson, Garden (London 1871-1927) 25: 276 (1884). *Magnolia liliiflora* var. *nigra* (G.Nicholson) Rehder in L.H.Bailey, Stand. Cycl. Hort. 4: 1968 (1916).

Magnolia × **loebneri** Kache, Gartenschönheit 1: 20 (1920). M. kobus × M. stellata. *Magnolia kobus* var. *loebneri* (Kache) Spongberg, J. Arnold Arbor. 57: 297 (1976).
Cult. 00 CUL. Nanophan. or phan.

Magnolia macklottii (Korth.) Dandy, Bull. Misc. Inform. Kew 1927: 263 (1927).
Malesia. 42 BOR JAW MLY SUM. Nanophan. or phan.
* *Manglietia macklottii* Korth., Ned. Kruidk. Arch. 2(2): 97 (1851).

var. **beccariana** (A.Agostini) Noot., Blumea 32: 348 (1987).
Sumatra, Pen. Malaysia. 42 MLY SUM. Nanophan. or phan.
* *Michelia beccariana* A.Agostini, Atti Reale Accad. Fisiocrit. Siena, X, 1: 184 (1926).
Magnolia aequinoctialis Dandy, Bull. Misc. Inform. Kew 1928: 185 (1928).

var. **macklottii**
W. Sumatra, W. Java, Borneo (Sabah, Tawau). 42 BOR JAW SUM. Nanophan. or phan.
Magnolia javanica Koord. & Valeton, Meded. Lands Plantentuin 17: 315 (1896).

Magnolia macrophylla Michx., Fl. Bor.-Amer. 1: 327 (1803).
E. U.S.A., Mexico, Caribbean. 75 OHI WVA? 78 ALA ARK GEO KTY LOU MSI NCA SCA TEN VRG 79 MXE MXS 81 CUB DOM HAI PUE. Phan.

subsp. **ashei** (Weath.) Spongberg, J. Arnold Arbor. 57: 268 (1976).
NW. Florida. 78 FLA. Phan.
* *Magnolia ashei* Weath., Rhodora 28: 35 (1926). *Magnolia macrophylla* var. *ashei* (Weath.) D.L.Johnson, Baileya 23: 55 (1989).

subsp. **macrophylla**
E. U.S.A. 75 OHI WVA? 78 ALA ARK GEO KTY LOU MSI NCA SCA TEN VRG. Phan.
Magnolia michauxiana DC., Syst. Nat. 1: 455 (1817).
Magnolia pilosissima P.Parm., Bull. Sc. France Belgique 27: 196, 254 (1896).

Magnolia magnifolia (Lozano) Govaerts in D.G.Frodin & R.Govaerts, World Checklist Bibliogr. Magnoliaceae: 71 (1996).
Colombia (Choco). 83 CLM. Phan.
* *Dugandiodendron magnifolium* Lozano, Fl. Colombia 1: 37 (1983).

Magnolia mahechae (Lozano) Govaerts in D.G.Frodin & R.Govaerts, World Checklist Bibliogr. Magnoliaceae: 71 (1996).
Colombia (Valle). 83 CLM. Phan.
* *Dugandiodendron mahechae* Lozano, Caldasia 11(53): 33 (1975).

Magnolia maingayi King, J. Asiat. Soc. Bengal 58(2): 369 (1889).
Pen. Malaysia (incl. Singapore), Borneo (Sarawak, Sabah). 42 BOR MLY. Nanophan. or phan.

Magnolia mariusjacobsia Noot., Blumea 32: 381 (1987).
Borneo (Sarawak: Kapit). 42 BOR. Phan.

Magnolia mexicana DC., Syst. Nat. 1: 451 (1817). *Talauma mexicana* (DC.) G.Don, Gen. Hist. 1: 85 (1831).
Mexico, Guatemala, Honduras. 79 MXC MXE MXG MXS SMX 80 GUA HON. Phan.
Talauma macrocarpa Zucc., Abh. Math.-Phys. Cl. Königl. Bayer. Akad. Wiss. 2: 369 (1836).

Magnolia minor (Urb.) Govaerts in D.G.Frodin & R.Govaerts, World Checklist Bibliogr. Magnoliaceae: 71 (1996).
E. Cuba. 81 CUB. Phan.
* *Talauma minor* Urb., Symb. Antill. 7: 222 (1912). *Svenhedinia minor* (Urb.) Urb., Repert. Spec. Nov. Regni Veg. 24: 3 (1927).
Talauma orbiculata Britton & Wilson, Bull. Torrey Bot. Club 50: 37 (1923). *Talauma minor* subsp. *orbiculata* (Britton & Wilson) Borhidi, Acta Bot. Acad. Sci. Hung. 17: 7 (1971 publ. 1972).
Svenhedinia truncata Moldenke, Phytologia 2: 142 (1946). *Talauma truncata* (Moldenke) R.A.Howard, Bull. Torrey Bot. Club 75: 357 (1948).
Talauma minor var. *oblongifolia* Léon, Contrib. Ocas. Mus. Hist. Nat. Calegio "De La Salle" 9: 5 (1950). *Talauma minor* subsp. *oblongifolia* (Léon) Borhidi, Acta Bot. Acad. Sci. Hung. 17: 6 (1971 publ. 1972). *Talauma oblongifolia* (Léon) Bisse, Repert. Spec. Nov. Regni Veg. 85: 589 (1974).
Talauma ophiticola Bisse, Repert. Spec. Nov. Regni Veg. 85: 589 (1974).

Magnolia morii (Lozano) Govaerts in D.G.Frodin & R.Govaerts, World Checklist Bibliogr. Magnoliaceae: 71 (1996).
Panama (Panama). 80 PAN. Phan.
* *Talauma morii* Lozano, Dugandiodendron Talauma Neotróp.: 113 (1994).

Magnolia multiflora M.C.Wang & C.L.Min, Acta Bot. Boreal.-Occid. Sin. 12: 85 (1992).
SW. China. 36 CHC. Phan.

Magnolia nana Dandy, J. Bot. 68: 207 (1930).
Vietnam. 41 VIE. Nanophan.

Magnolia narinensis (Lozano) Govaerts in D.G.Frodin & R.Govaerts, World Checklist Bibliogr. Magnoliaceae: 71 (1996).
Colombia (Narierts in D.G.Frodin
* *Talauma narinensis* Lozano, Caldasia 12(58): 286 (1978).

Magnolia neillii (Lozano) Govaerts in D.G.Frodin & R.Govaerts, World Checklist Bibliogr. Magnoliaceae: 71 (1996).
Ecuador (Napo). 83 ECU. Phan.
* *Talauma neillii* Lozano, Dugandiodendron Talauma Neotróp.: 71 (1994).

Magnolia nitida W.W.Sm., Notes Roy. Bot. Gard. Edinburgh 12: 212 (1920). *Parakmeria nitida* (W.W.Sm.) Y.W.Law in W.C.Cheng, Sylva Sinica 1: 472 (1983). *Parakmeria nitida* (W.W.Sm.) Y.W.Law, Acta Phytotax. Sin. 34: 91 (1996).
NE. Burma, SE. Tibet, Vietnam, China. 36 CHC CHH CHS CHT 41 BMA VIE. Phan.

var. **lotungensis** (Chun & C.H.Tsoong) B.L.Chen & Noot., Ann. Missouri Bot. Gard. 80: 1016 (1994).
China (Guangdong, Guangxi, Guizhou, Hunan, Zhejiang), Hainan. 36 CHC CHH CHS. Phan.
* *Magnolia lotungensis* Chun & C.H.Tsoong, Acta Phytotax. Sin. 8: 285 (1963). *Parakmera lotungensis* (Chun & C.H.Tsoong) Y.W.Law, Acta Phytotax. Sin. 34: 91 (1996).

var. **nitida**
NE. Burma, SE. Tibet, SW. China (Yunnan). 36 CHC CHT 41 BMA. Phan.
Parakmeria yunnanensis Hu, Acta Phytotax. Sin. 1: 2 (1951). *Magnolia yunnanensis* (Hu) Noot., Blumea 31: 88 (1985).
Parakmeria nitida (W.W.Sm.) Y.W.Law in W.C.Cheng, Sylva Sinica 1: 472 (1983).
Parakmeria nitida (W.W.Sm.) Y.W.Law, Acta Phytotax. Sin. 34: 91 (1996).

var. **robusta** B.L.Chen & Noot., Ann. Missouri Bot. Gard. 40: 1016 (1993).
China (Guangxi), Vietnam. 36 CHS 41 VIE. Phan.

Magnolia obovata Thunb., Trans. Linn. Soc. London 2: 336 (1794). *Yulania japonica* var. *obovata* (Thunb.) P.Parm., Bull. Sc. France Belgique 27: 258 (1896).
  S. Kuril Is., Japan, Nansei-shoto. 31 KUR 38 JAP NNS. Phan.
    Magnolia glauca Thunb., Fl. Jap.: 236 (1784).
    Magnolia hoonokii Siebold, Verh. Batav. Genootsch. Kunsten 12: 50 (1830).
    Liriodendron liliiflorum Steud., Nomencl. Bot., ed. 2, 2: 55 (1841).
    Magnolia hypoleuca Siebold & Zucc., Abh. Math.-Phys. Cl. Königl. Bayer. Akad. Wiss. 4: 187 (1845).
    Magnolia hypoleuca var. *concolor* Siebold & Zucc., Abh. Math.-Phys. Cl. Königl. Bayer. Akad. Wiss. 4: 187 (1845).
    Magnolia honogi P.Parm., Bull. Sc. France Belgique 27: 195, 254 (1896), orth. var.

Magnolia officinalis Rehder & E.H.Wilson in C.S.Sargent, Pl. Wilson. 1: 391 (1913).
  Tibet, China (Anhui, Fujian, Guangdong, Guangxi, Guizhou, Hunan, Hubei, Jiangxi, Zhejiang, Shaanxi, Sichuan). 36 CHC CHN CHS CHT. Phan.

  var. **officinalis**
    Tibet, China (Anhui, Guangxi, Guizhou, Hunan, Hubei, Jiangxi, Zhejiang, Shaanxi, Sichuan). 36 CHC CHN CHS CHT. Phan.
    Magnolia officinalis var. *pubescens* C.Y.Deng, J. Nanjing Inst. Forest. 1986: 145 (1986).

  var. **biloba** Rehder & E.H.Wilson in C.S.Sargent, Pl. Wilson. 1: 392 (1913).
    China (Anhui, Fujian, Guangdong, Guangxi, Hunan, Jiangxi, Zhejiang). 36 CHS. Phan.
    Magnolia biloba (Rehder & E.H.Wilson) W.C.Cheng & Y.W.Law, Icon. Cormophyt. Sin. 1: 787 (1972).
    Magnolia officinalis subsp. *biloba* (Rehder & E.H.Wilson) W.C.Cheng & Y.W.Law in W.C.Cheng, Sylva Sinica 1: 449 (1983).

Magnolia omeiensis (Hu & C.Y.Cheng) Dandy in S.Nilsson, World Pollen Spore Fl. 3(Magnoliaceae): 5 (1974).
  SW. China (Sichuan: Emei Shan, Guizhou ?). 36 CHC. Phan.
    * *Parakmeria omeiensis* W.C.Cheng, Acta Phytotax. Sin. 1: 2 (1951).

Magnolia ovata (A.St.-Hil.) Spreng., Syst. Veg. 4(2): 217 (1827).
  Brazil (Goiás, Distrito Federal, Minas Gerais). 84 BZC BZL. Phan.
    * *Talauma ovata* A.St.-Hil., Fl. Bras. Merid. 1: 26 (1824).
    Talauma dubia Eichler in C.F.P.von Martius, Fl. Bras. 13(1): 126 (1864).

Magnolia pacifica Vázquez, Brittonia 46: 10 (1994).
  Mexico. 79 MXE MXN MXS. Phan.

  subsp. **pacifica**
    Mexico (Jalisco, Nayarit). 79 MXS. Phan.

  subsp. **pugana** H.H.Iltis & Vázquez, Brittonia 46: 14 (1994).
    Mexico (Jalisco, Zacatecas). 79 MXS. Phan.

  subsp. **tarahumara** Vázquez, Brittonia 46: 14 (1994).
    Mexico (Sonora, Chihuahua, Sinaloa, Durango). 79 MXE MXN. Phan.

Magnolia pahangensis Noot., Blumea 32: 367 (1987).
  Pen. Malaysia (Pahang). 42 MLY. Phan.

Magnolia pallescens Urb. & Ekman, Ark. Bot. 23A(11): 10 (1931).
  W. Dominican Rep. 81 DOM. Phan.

Magnolia panamensis H.H.Iltis & Vázquez, Brittonia 46: 15 (1994).
  Panama. 80 PAN. Phan.

Magnolia pealiana King, Ann. Roy. Bot. Gard. (Calcutta) 3(2): 210 (1891). *Magnolia membranacea* var. *pealiana* (King) P.Parm., Bull. Sc. France Belgique 27: 200 (1896). *Michelia pealiana* (King) Finet & Gagnep., Bull. Soc. Bot. France 52(4): 42 (1906). Assam (Makum forest). 40 ASS. Phan.

Magnolia persuaveolens Dandy, Bull. Misc. Inform. Kew 1928: 186 (1928). *Talauma persuaveolens* (Dandy) Dandy, Taxon 21: 468 (1972). Borneo (Sabah: Mt. Kinabalu). 42 BOR. Nanophan.

> subsp. **persuaveolens**
> Borneo (Sabah: Mt. Kinabalu). 42 BOR. Nanophan.

> var. **pubescens** Noot., Blumea 32: 379 (1987).
> Borneo (Sabah: Mt. Kinabalu). 42 BOR. Nanophan. or phan.

> subsp. **rigida** Noot., Blumea 32: 379 (1987).
> Borneo (Sabah: Mt. Kinabalu). 42 BOR. Nanophan. or phan.
> *Magnolia persuaveolens* var. *rigida* Noot., Blumea 32: 379 (1987).

Magnolia phanerophlebia B.L.Chen, Acta Sci. Nat. Univ. Sunyatseni 1988: 107 (1988). SW. China (SE. Yunnan). 36 CHC. Nanophan.

Magnolia phaulanta Dandy ex Noot., Blumea 32: 359 (1987). Sulawesi. 42 SUL. Phan.

Magnolia poasana (Pittier) Dandy, Bull. Misc. Inform. Kew 1927: 263 (1927). Costa Rica, Panama. 80 COS PAN. Phan.
> * *Talauma poasana* Pittier, Contrib. U.S. Natl. Herb. 13: 93 (1910).

Magnolia poilanei Dandy ex Gagnep. in P.H.Lecomte, Fl. Indo-Chine, Suppl. 1: 40 (1938). Vietnam. 41 VIE. Phan.

Magnolia polyhypsophylla (Lozano) Govaerts in D.G.Frodin & R.Govaerts, World Checklist Bibliogr. Magnoliaceae: 71 (1996). Colombia (Antioquia). 83 CLM. Phan.
> * *Talauma polyhypsophylla* Lozano, Fl. Colombia 1: 87 (1983).

Magnolia portoricensis Bello, Anales Soc. Esp. Hist. Nat. 10: 233 (1880). W. Puerto Rico. 81 PUE. Phan.
> *Magnolia patoricensis* P.Parm., Bull. Sc. France Belgique 27: 205, 269 (1896).

Magnolia ptaritepuiana Steyerm., Fieldiana, Bot. 28(1): 233 (1951). *Dugandiodendron ptaritepuianum* (Steyerm.) Lozano, Caldasia 11(53): 42 (1975). Venezuela (Bolivar). 82 VEN. Phan.
> *Magnolia roraimae* Steyerm., Fieldiana, Bot. 28(1): 234 (1951).

Magnolia pterocarpa Roxb., Pl. Coromandel 3: 62 (1820). *Sphenocarpus pterocarpus* (Roxb.) K.Koch, Hort. Dendrol. 5: 1 (1853). Nepal, Bhutan, Assam, Burma ?, Thailand ?. 40 ASS BHU NEP 41 BMA? THA?. Phan.
> *Liriodendron grandiflorum* Roxb., Fl. Ind. 2: 653 (1824). *Talauma roxburghii* G.Don, Gen. Hist. 1: 85 (1831).
> *Liriodendron indicum* Spreng., Syst. Veg. 2: 642 (1825).
> *Michelia macrophylla* D.Don, Prodr. Fl. Nepal.: 226 (1825).
> *Lirianthe grandiflora* Spach, Hist. Nat. Vég. 7: 485 (1839).
> *Magnolia sphenocarpa* Hook.f. & Thomson, Fl. Ind. 1: 78 (1855), sphalm. M. speciosa.

Magnolia pulgarensis Dandy, Bull. Misc. Inform. Kew 1928: 187 (1928). Philippines (Palawan: Mt. Pulgar). 42 PHI. Phan.

Magnolia rimachii (Lozano) Govaerts in D.G.Frodin & R.Govaerts, World Checklist
    Bibliogr. Magnoliaceae: 71 (1996).
    Ecuador (Napo), N. Peru. 83 ECU PER. Phan.
    * Talauma rimachii Lozano, Dugandiodendron Talauma Neotróp.: 105 (1994).

Magnolia rostrata W.W.Sm., Notes Roy. Bot. Gard. Edinburgh 12: 213 (1920).
    NE. Burma, Tibet, SW. China (Yunnan). 36 CHS CHT 41 BMA. Phan.

Magnolia salicifolia (Siebold & Zucc.) Maxim., Bull. Acad. Imp. Sc. Saint-Pétersbourg 17:
    418 (1872).
    Japan (Honshu, Shikoku, Kyushu). 38 JAP. Nanophan. or phan.
        Magnolia salicifolia var. *fasciata* Millais in ? *Magnolia salicifolia* f. *fasciata* (Millais)
            Rehder, Bibl. Cult. Trees: 181 (1849).
        * Buergeria salicifolia Siebold & Zucc., Abh. Math.-Phys. Cl. Königl. Bayer. Akad. Wiss. 4:
            187 (1845).
        Talauma salicifolia var. *concolor* Miq., Ann. Mus. Bot. Lugduno-Batavi 2: 258 (1866).
            Magnolia salicifolia var. *concolor* (Miq.) Maxim., Bull. Acad. Imp. Sc. Saint-Pétersbourg
            17: 418 (1872).
        Magnolia famasiha P.Parm., Bull. Sc. France Belgique 27: 200, 259 (1896).
        Magnolia proctoriana Rehder, J. Arnold Arbor. 20: 412 (1939).
        Magnolia kewensis Pearce, Gard. Chron. 132: 154 (1952), nom. nud.
        Magnolia slavinii Harkn., Natl. Hort. Mag. 33: 118 (1954).

Magnolia sambuensis (Pittier) Govaerts in D.G.Frodin & R.Govaerts, World Checklist
    Bibliogr. Magnoliaceae: 72 (1996).
    Panama, Colombia (Antioquia, Choco). 80 PAN 83 CLM. Phan.
    * Talauma sambuensis Pittier, Contrib. U.S. Natl. Herb. 20: 105 (1918).

Magnolia santanderiana (Lozano) Govaerts in D.G.Frodin & R.Govaerts, World Checklist
    Bibliogr. Magnoliaceae: 72 (1996).
    Colombia (Santander). 83 CLM. Phan.
    * Talauma santanderiana Lozano, Fl. Colombia 1: 95 (1983).

Magnolia sarawakensis (A.Agostini) Noot., Blumea 32: 380 (1987).
    Borneo (Sarawak, Sabah, E. & W. Kalimantan). 42 BOR. Phan.
    * Talauma sarawakensis A.Agostini, Atti Reale Accad. Fisiocrit. Siena, X, 1: 190 (1926).
        Talauma intonsa Dandy, Bull. Misc. Inform. Kew 1928: 191 (1928).

Magnolia sargentiana Rehder & E.H.Wilson in C.S.Sargent, Pl. Wilson. 1: 398 (1913).
    SW. China (Sichuan, Yunnan). 36 CHC. Phan.
        Magnolia conspicua var. *emarginata* Finet & Gagnep., Bull. Soc. Bot. France 52(4): 38
            (1905). *Magnolia denudata* var. *emarginata* (Finet & Gagnep.) Pamp., Bull. Soc. Tosc.
            Ortic. 20: 200 (1915). *Magnolia emarginata* (Finet & Gagnep.) W.C.Cheng, J. Bot. Soc.
            China 1(3): 298 (1934).
        Magnolia sargentiana var. *robusta* Rehder & E.H.Wilson, Pl. Wilson. 1: 399 (1913).

Magnolia schiedeana Schltl., Bot. Zeit. (Berlin) 22: 144 (1864).
    Mexico (Sinaloa & Nayarit to Veracruz). 79 MXE MXG MXN MXS. Phan.

Magnolia sellowiana (A.St.-Hil.) Govaerts in D.G.Frodin & R.Govaerts, World Checklist
    Bibliogr. Magnoliaceae: 72 (1996).
    Brazil (Minas Gerais, São Paulo, Rio de Janeiro). 84 BZL. Phan.
    * Talauma sellowiana A.St.-Hil., Fl. Bras. Merid. 1: 26 (1824).
        Magnolia selloi Spreng., Syst. Veg. 4(2): 216 (1827). *Talauma selloi* (Spreng.) Steud.,
            Nomencl. Bot., ed. 2, 2: 90 (1841).
        Talauma fragrantissima Hook., Hooker's Icon. Pl. 3: 208-212 (1839).

Magnolia sharpii V.V.Miranda, Anales Inst. Biol. Univ. Nac. Mexico 26: 79 (1955).
S. Mexico (Chiapas). 79 SMX. Phan.

Magnolia sieboldii K.Koch, Hort. Dendrol. 4: 11 (1853)
Korea, China (Liaoning, Anhui, Guangxi, Guizhou, Yunnan, Sichuan).36 CHC CHM CHS
38 KOR. Nanophan. or phan.

subsp. **sieboldii**
Korea, China (Liaoning, Anhui, Guangxi, Guizhou, Yunnan, Sichuan).36 CHC CHM
CHS 38 KOR. Nanophan. or phan.
Magnolia parviflora Siebold & Zucc., Abh. Math.-Phys. Cl. Königl. Bayer. Akad. Wiss. 4:
187 (1845).
Magnolia oyama Kort, Rev. Hort. Belge Etrangère 31: 258 (1905).
Magnolia verecunda Koidz., Bot. Mag. (Tokyo) 40: 339 (1926).

subsp. **japonica** K.Ueda, Acta Phytotax. Geobot. 31: 121 (1980).
Japan (C. & W. Honshu, Shikoku, Kyushu), China (Anhui, Guangxi) ?. 36 CHS? 38 JAP.
Nanophan.

subsp. **sinensis** (Rehder & E.H.Wilson) Spongberg, J. Arnold Arbor. 57: 279 (1976).
China (NW. Sichuan). 36 CHC. Phan.
* *Magnolia globosa* var. *sinensis* Rehder & E.H.Wilson in C.S.Sargent, Pl. Wilson. 1: 393
(1913). *Magnolia sinensis* (Rehder & E.H.Wilson) Stapf, Bot. Mag.: 9004 (1924).

Magnolia silvioi (Lozano) Govaerts in D.G.Frodin & R.Govaerts, World Checklist Bibliogr.
Magnoliaceae: 72 (1996).
Colombia (Antioquia). 83 CLM. Phan.
* *Talauma silvioi* Lozano, Fl. Colombia 1: 98 (1983).

Magnolia sororum Seibert, Ann. Missouri Bot. Gard. 25: 828 (1938).
Costa Rica, Panama. 80 COS PAN. Phan.

subsp. **lutea** Vázquez, Brittonia 46: 18 (1994).
Costa Rica. 80 COS. Phan.

subsp. **sororum**
Panama. 80 PAN. Phan.

Magnolia × soulangeana Soul.-Bod., Trans. Linn. Soc. Paris 1826: 269 (1826). M. denudata
× M. liliifera. *Magnolia obovata* var. *soulangeana* (Soul.-Bod.) Ser. in ? *Magnolia conspicua*
var. *soulangeana* (Soul.-Bod.) Loudon in ? *Magnolia yulan* var. *soulangeana* (Soul.-Bod.)
Lindl. in S.T.Edwards, Bot. Reg. 14: 1164 (1828).
Cult. 00 CUL. Nanophan. or phan.
Magnolia × soulangeana var. *brozzinii* auct. in ? *Magnolia × brozzonii* (auct.) Millais,
Magnolias: 56, 87 (1927).
Yulania × lenneana Lem., Ill. Hort. 1: 37 (1854). *Magnolia × lenneana* (Lem.) Koehne,
Deut. Dendrol.: 146 (1893).
Magnolia × lennei Van Houtte, Fl. Serres 16: 1693 (1867).

Magnolia splendens Urb., Symb. Antill. 1: 306 (1899). *Talauma splendens* (Urb.)
McLaughlin, Trop. Woods 34: 36 (1933). *Talauma mutabilis* var. *splendens* (Urb.) Urb. ex
McLaughlin, Trop. Woods 34: 36 (1933).
C. Puerto Rico. 81 PUE. Phan.

Magnolia sprengeri Pamp., Nuovo Giorn. Bot. Ital. n.s. 22: 295 (1915).
China (Henan, Hubei, Guizhou, Sichuan, Yunnan). 36 CHC CHS. Phan.
Magnolia conspicua var. *purpurascens* Maxim., Bull. Acad. Imp. Sc. Saint-Pétersbourgh
17: 419 (1872). *Magnolia denudata* var. *purpurascens* (Maxim.) Rehder & E.H.Wilson
in C.S.Sargent, Pl. Wilson. 1: 401 (1913). *Magnolia purpurascens* (Maxim.) Makino, J.

Jap. Bot. 6(4): 8 (1929). *Magnolia heptapeta* f. *purpurascens* (Maxim.) H.Ohba, J. Jap. Bot. 55: 190 (1980).

Magnolia denudata var. *elongata* Rehder & E.H.Wilson in C.S.Sargent, Pl. Wilson. 1: 402 (1913). *Magnolia elongata* (Rehder & E.H.Wilson) Millais, Magnolias: 59 (1927). *Magnolia sprengeri* var. *elongata* (Rehder & E.H.Wilson) Johnstone, Asiatic Magnolias Cult.: 87 (1955).

Magnolia diva Stapf ex Dandy in J.G.Millais, Magnolias: 51, 120 (1927). *Magnolia sprengeri* var. *diva* (Stapf ex Dandy) Stapf, Bot. Mag. 153: 9116 (1927).

Magnolia stellata (Siebold & Zucc.) Maxim., Bull. Acad. Imp. Sc. Saint-Pétersbourg 17: 418 (1872).

Japan (SC. Honshu: Tokaido). 38 JAP. Nanophan.

Magnolia kobus f. *stellata* (Siebold & Zucc.) Maxim. in ?

* Buergeria stellata Siebold & Zucc., Abh. Math.-Phys. Cl. Königl. Bayer. Akad. Wiss. 4: 186 (1845). *Magnolia kobus* f. *stellata* (Siebold & Zucc.) Maxim. in ? *Talauma stellata* (Siebold & Zucc.) Miq., Ann. Mus. Bot. Lugduno-Batavi 2: 257 (1866). *Magnolia kobus* var. *stellata* (Siebold & Zucc.) Blackburn, Amatores Herb. 17: 2 (1955).

Magnolia simii Siebold ex Miq., Ann. Mus. Bot. Lugduno-Batavi 2: 257 (1866).

Magnolia halleana auct., Fl. Mag. (London): 309 (1878).

Magnolia stellata var. *keiskei* Makino, Bot. Mag. (Tokyo) 26: 82 (1912). *Magnolia keiskei* (Makino) Ihrig, Arbor. Bull., Washington 11(2): 33 (1948).

Magnolia stellata var. *rosea* Veitch ex Hu, Man. Econ. Pl. 1: 383 (1912). *Magnolia rosea* (Veitch ex Hu) Ihrig, Arbor. Bull., Washington 11(2): 34 (1948). *Magnolia kobus* f. *rosea* (Veitch ex Hu) Blackburn, Popular Gard. (Albany) 5(3): 73 (1954).

Magnolia sinostellata P.L.Chiu & Z.H.Chen, Acta Phytotax. Sin. 27: 79 (1989).

Magnolia striatifolia Little, Phytologia 18: 198 (1969). *Dugandiodendron striatifolium* (Little) Lozano, Caldasia 11(53): 44 (1975).

Colombia (Nariño), Ecuador (Esmeraldas). 83 CLM ECU. Phan.

Magnolia tamaulipana Vázquez, Brittonia 46: 18 (1994).

Mexico (Tamaulipas, Nuevo Leon). 79 MXE. Phan.

Magnolia × thompsoniana de Vos, Ned. Fl. & Pomon.: 131 (1876). M. tripetala × M. virginiana. Cult. 00 CUL.

Magnolia glauca var. *major* Sims, Bot. Mag. t. 2164 (1820).

* Magnolia glauca var. *thompsoniana* Loudon, Arbor. Frutic. Brit. 1: 267 (1854), nom. illeg.

Magnolia tripetala (L.) L., Syst. Nat. ed. 10: 1082 (1759).

E. U.S.A. to Oklahoma. 74 OKL 75 INI OHI PEN WVA 78 ALA ARK FLA GEO KTY MRY MSI NCA SCA TEN VRG. Phan.

* Magnolia virginiana var. *tripetala* L., Sp. Pl.: 536 (1753).

Magnolia umbrella Desr. in J.B.A.M.de Lamarck, Encycl. 3: 673 (1792).

Magnolia frondosa Salisb., Prodr. Stirp. Chap. Allerton: 379 (1796).

Magnolia umbellata Steud., Nomencl. Bot., ed. 2, 2: 90 (1841).

Magnolia urraoense (Lozano) Govaerts in D.G.Frodin & R.Govaerts, World Checklist Bibliogr. Magnoliaceae: 72 (1996).

Colombia (Antioquia). 83 CLM. Phan.

* Dugandiodendron urraoense Lozano, Fl. Colombia 1: 42 (1983).

Magnolia uvariifolia Dandy ex Noot., Blumea 32: 358 (1987).

Borneo. 42 BOR. Phan.

Magnolia × veitchii Bean, J. Roy. Hort. Soc. 46: 321 (1921). M. campbellii × M. denudata. Cult. 00 CUL. Phan.

Magnolia venezuelensis (Lozano) Govaerts in D.G.Frodin & R.Govaerts, World Checklist Bibliogr. Magnoliaceae: 72 (1996).
N. Venezuela. 82 VEN. Phan.
* *Talauma venezuelensis* Lozano, Revista Acad. Colomb. Ci. Exact. 17: 78 (1990).

Magnolia villosa (Miq.) H.Keng, Gard. Bull. Singapore 31: 129 (1978).
Pen. Malaysia, Sumatra (incl. Lingga), Borneo (Sabah). 42 BOR MLY SUM. Phan.
* *Talauma villosa* Miq., Fl. Ned. Ind. Eerste Bijv.: 366 (1861). *Talauma rabaniana* var. *villosa* (Miq.) P.Parm., Bull. Sc. France Belgique 27: 271 (1896).
Talauma lanigera Hook.f. & Thomson in J.D.Hooker, Fl. Brit. Ind. 1: 40 (1872). *Magnolia lanigera* (Hook.f. & Thomson) H.J.Chowdhery & P.Daniel, Indian J. Forest. 4: 64 (1981).

Magnolia virginiana L., Sp. Pl.: 535 (1753).
E. U.S.A. to Texas. 75 MAS NWJ NWY PEN WDC 77 TEX 78 ALA ARK DEL FLA GEO LOU MRY MSI NCA SCA TEN. Nanophan. or phan.
Magnolia virginiana var. *glauca* L. in ? *Magnolia glauca* (L.) L., Syst. Nat. ed. 10: 1082 (1759).
Magnolia glauca var. *longifolia* Aiton, Hortus Kew. 2: 251 (1789).
Magnolia virginiana var. *longifolia* Aiton, Hort. Kew. 2: 251 (1789).
Magnolia fragrans Salisb., Prodr. Stirp. Chap. Allerton: 379 (1796).
Magnolia virginiana var. *pumila* Nutt., Amer. J. Sci. 5: 295 (1822).
Magnolia glauca var. *argentea* DC., Prodr. 1: 80 (1824).
Magnolia burchelliana Steud., Nomencl. Bot., ed. 2, 2: 89 (1841).
Magnolia gordoniana Steud., Nomencl. Bot., ed. 2, 2: 90 (1841).
Magnolia latifolia Aiton ex Dippel, Handb. Laubholzk. 3: 145 (1893).
Magnolia virginiana var. *australis* Sarg., Bot. Gaz. 67: 231 (1919). *Magnolia australis* (Sarg.) Ashe, Torreya 31: 39 (1931). *Magnolia virginiana* subsp. *australis* (Sarg.) E.Murray, Kalmia 11: 2 (1981).
Magnolia major Millais, Magnolias: 66 (1927).
Magnolia virginiana var. *parva* Ashe, Bull. Torrey Bot. Club 55: 404 (1928). *Magnolia australis* var. *parva* (Ashe) Ashe, Torreya 31: 39 (1931).

Magnolia virolinensis (Lozano) Govaerts in D.G.Frodin & R.Govaerts, World Checklist Bibliogr. Magnoliaceae: 72 (1996).
Colombia (Santander). 83 CLM. Phan.
* *Talauma virolinensis* Lozano, Fl. Colombia 1: 102 (1983).

Magnolia × **wiesneri** Carr., Rev. Hort. 62: 406 (1890). M. obovata × M. sieboldii.
Cult. 00 CUL.
Magnolia × *watsonii* Hook.f., Bot. Mag.: 7157 (1891).

Magnolia wilsonii (Finet & Gagnep.) Rehder in C.S.Sargent, Pl Wilson. 1: 395 (1913).
SW. China (Sichuan, N. Yunnan, Guizhou). 36 CHC. Nanophan. or phan.
Magnolia liliifera var. *taliensis* (W.W.Sm.) Pamp. in ?
* *Magnolia parviflora* var. *wilsonii* Finet & Gagnep., Bull. Soc. Bot. France 52(4): 39 (1905).
Magnolia nicholsoniana Rehder & E.H.Wilson in C.S.Sargent, Pl. Wilson. 1: 394 (1913). *Magnolia wilsonii* f. *nicholsoniana* (Rehder & E.H.Wilson) Rehder, J. Arnold Arbor. 20: 91 (1939).
Magnolia taliensis W.W.Sm., Notes Roy. Bot. Gard. Edinburgh 8: 341 (1915). *Magnolia liliifera* var. *taliensis* (W.W.Sm.) Pamp. in ? *Magnolia wilsonii* f. *taliensis* (W.W.Sm.) Rehder, Man. Cult. Trees: 249 (1940).
Magnolia highdownensis Dandy, J. Roy. Hort. Soc. 75: 159 (1950).

Magnolia wolfii (Lozano) Govaerts in D.G.Frodin & R.Govaerts, World Checklist Bibliogr. Magnoliaceae: 72 (1996).
  Colombia (Risazalda). 83 CLM. Phan.
  * *Talauma wolfii* Lozano, Dugandiodendron Talauma Neotróp.: 90 (1994).

Magnolia yarumalense (Lozano) Govaerts in D.G.Frodin & R.Govaerts, World Checklist Bibliogr. Magnoliaceae: 72 (1996).
  Colombia (Antioquia). 83 CLM. Phan.
  * *Dugandiodendron yarumalense* Lozano, Fl. Colombia 1: 46 (1983).

Magnolia yoroconte Dandy, J. Bot. 68: 147 (1930).
  Mexico (Veracruz, Chiapas), Belize, N. Guatemala, Honduras. 79 MXG SMX 80 BLZ GUA HON. Phan.

Magnolia zenii W.C.Cheng, Contrib. Biol. Lab. Chin. Assoc. Advancem. Sci., Sect. Bot. 8: 291 (1933).
  China (Henan, Jiangsu). 36 CHS. Phan.
    Magnolia elliptilimba Y.W.Law & Z.Y.Gao, Bull. Bot. Res., Harbin 4: 190 (1984).

Synonyms:

Magnolia acuminata var. *alabamensis* Ashe === **Magnolia acuminata** var. **subcordata** (Spach) Dandy

Magnolia acuminata f. *aurea* (Ashe) Hardin === **Magnolia acuminata** (L.) L. var. **acuminata**

Magnolia acuminata var. *aurea* (Ashe) Ashe === **Magnolia acuminata** (L.) L. var. **acuminata**

Magnolia acuminata var. *cordata* (Michx.) Sarg. === **Magnolia acuminata** var. **subcordata** (Spach) Dandy

Magnolia acuminata subsp. *cordata* (Michx.) E.Murray === **Magnolia acuminata** var. **subcordata** (Spach) Dandy

Magnolia acuminata var. *decandollei* (Savi) DC. === **Magnolia acuminata** (L.) L. var. **acuminata**

Magnolia acuminata var. *ludoviciana* Sarg. === **Magnolia acuminata** (L.) L. var. **acuminata**

Magnolia acuminata subsp. *ozarkensis* (Ashe) E.Murray === **Magnolia acuminata** var. **ozarkensis** Ashe

Magnolia aequinoctialis Dandy === **Magnolia macklottii** var. **beccariana** (A.Agostini) Noot.

Magnolia alexandrina Steud. === **Magnolia denudata** Desr.

Magnolia andamanica (King) D.C.S.Raju & M.P.Nayar === **Magnolia liliifera** (L.) Baill. var. **liliifera**

Magnolia andamanica King === **Magnolia liliifera** (L.) Baill. var. **liliifera**

Magnolia angatensis Blanco === **Magnolia liliifera** var. **angatensis** (Blanco) Govaerts

Magnolia angustifolia Millais === **Magnolia grandiflora** L.

Magnolia annamensis var. *affinis* Gagnep. === **Magnolia annamensis** Dandy

Magnolia annonifolia Salisb. === **Michelia figo** (Lour.) Spreng. var. **figo**

Magnolia ashei Weath. === **Magnolia macrophylla** subsp. **ashei** (Weath.) Spongberg

Magnolia atropurpurea Steud. === **Magnolia liliiflora** Desr.

Magnolia aulacosperma Rehder & E.H.Wilson === **Magnolia biondii** Pamp.

Magnolia auricularis Salisb. === **Magnolia fraseri** Walter var. **fraseri**

Magnolia auriculata Desr. === **Magnolia fraseri** Walter var. **fraseri**

Magnolia auriculata var. *pyramidata* (Bartram) Nutt. === **Magnolia fraseri** var. **pyramidata** (Bartram) Pamp.

Magnolia australis (Sarg.) Ashe === **Magnolia virginiana** L.

Magnolia australis var. *parva* (Ashe) Ashe === **Magnolia virginiana** L.

Magnolia axilliflora T.B.Chao & al. === **Magnolia biondii** Pamp.

Magnolia axilliflora var. *alba* T.B.Chao & al. === **Magnolia biondii** Pamp.

Magnolia axilliflora var. *multitepala* T.B.Chao & al. === **Magnolia biondii** Pamp.

Magnolia baillonii Pierre === **Michelia baillonii** (Pierre) Finet & Gagnep.

Magnolia balansae A.DC. === **Michelia balansae** (A.DC.) Dandy

Magnolia betongensis (Craib) H.Keng === **Magnolia liliifera** var. **obovata** (Korth.) Govaerts

Magnolia biloba (Rehder & E.H.Wilson) W.C.Cheng & Y.W.Law === **Magnolia officinalis** var. **biloba** Rehder & E.H.Wilson

Magnolia biondii var. *flava* T.B.Chao & al. === **Magnolia biondii** Pamp.

Magnolia biondii var. *ovata* T.B.Chao & T.X.Zhang === **Magnolia biondii** Pamp.

Magnolia biondii var. *parvialabastra* T.B.Chao & al. === **Magnolia biondii** Pamp.

Magnolia biondii var. *planities* T.B.Chao & T.Z.Qiao === **Magnolia biondii** Pamp.

Magnolia biondii f. *purpurascens* Y.W.Law & Z.Y.Gao === **Magnolia biondii** Pamp.

Magnolia biondii var. *purpurea* T.B.Chao & Y.C.Qiao === **Magnolia biondii** Pamp.

Magnolia biondii var. *tatitepala* T.B.Chao & J.T.Gao === **Magnolia biondii** Pamp.

Magnolia blaoensis (Gagnep.) Dandy === **Manglietia blaoensis** Gagnep.

Magnolia blumei Prantl === **Manglietia glauca** Blume var. **glauca**

Magnolia borealis (Sarg.) Kudô === **Magnolia kobus** DC.

Magnolia × *brozzonii* (auct.) Millais === **Magnolia** × **soulangeana** Soul.-Bod.

Magnolia burchelliana Steud. === **Magnolia virginiana** L.

Magnolia cacuminicola Bisse === **Magnolia cubensis** subsp. **cacuminicola** (Bisse) G.Klotz

Magnolia campbellii var. *alba* Tresder === **Magnolia campbellii** Hook.f. & Thomson

Magnolia campbellii subsp. *mollicomata* (W.W.Sm.) Johnstone === **Magnolia campbellii** Hook.f. & Thomson

Magnolia campbellii var. *mollicomata* (W.W.Sm.) F.S.Ward === **Magnolia campbellii** Hook.f. & Thomson

Magnolia candollei Link === **Magnolia acuminata** (L.) L. var. **acuminata**

Magnolia candollei (Blume) H.Keng === **Magnolia liliifera** (L.) Baill. var. **liliifera**

Magnolia candollei var. *angatensis* (Blanco) Noot. === **Magnolia liliifera** var. **angatensis** (Blanco) Govaerts

Magnolia candollei var. *beccarii* (Ridl.) Noot. === **Magnolia liliifera** var. **beccarii** (Ridl.) Govaerts

Magnolia candollei var. *obovata* (Korth.) Noot. === **Magnolia liliifera** var. **obovata** (Korth.) Govaerts

Magnolia candollei var. *singapurensis* (Ridl.) Noot. === **Magnolia liliifera** var. **singapurensis** (Ridl.) Govaerts

Magnolia caveana (Hook.f. & Thomson) D.C.S.Raju & M.P.Nayar === **Manglietia caveana** Hook.f. & Thomson

Magnolia champaca (L.) Baill. ex Pierre === **Michelia champaca** L.

Magnolia champacifolia Dandy ex Gagnep. === **Magnolia albosericea** Chun & C.H.Tsoong

Magnolia citriodora Steud. === **Magnolia denudata** Desr.

Magnolia compressa Maxim. === **Michelia compressa** (Maxim.) Sarg.

Magnolia conspicua Salisb. === **Magnolia denudata** Desr.

Magnolia conspicua var. *emarginata* Finet & Gagnep. === **Magnolia sargentiana** Rehder & E.H.Wilson

Magnolia conspicua var. *fargesii* Finet & Gagnep. === **Magnolia biondii** Pamp.

Magnolia conspicua var. *purpurascens* Rehder & E.H.Wilson === **Magnolia denudata** Desr.

Magnolia conspicua var. *purpurascens* Maxim. === **Magnolia sprengeri** Pamp.

Magnolia conspicua var. *rosea* Veits === **Magnolia denudata** Desr.

Magnolia conspicua var. *soulangeana* (Soul.-Bod.) Loudon === **Magnolia** × **soulangeana** Soul.-Bod.

Magnolia cordata Michx. === **Magnolia acuminata** var. **subcordata** (Spach) Dandy

Magnolia craibiana Dandy === **Magnolia liliifera** (L.) Baill. var. **liliifera**

Magnolia cubensis subsp. *acunae* Imkhan. === **Magnolia cubensis** Urb. subsp. **cubensis**

Magnolia cubensis var. *baracoensis* Imkhan. === **Magnolia cubensis** Urb. subsp. **cubensis**

Magnolia cyathiformis Rinz ex K.Koch === **Magnolia denudata** Desr.

Magnolia dandyi Gagnep. === **Manglietia dandyi** (Gagnep.) Dandy

Magnolia decandollei Savi === **Magnolia acuminata** (L.) L. var. **acuminata**

Magnolia denudata var. *angustitepala* T.B.Chao & Z.S.Chun === **Magnolia denudata** Desr.

Magnolia denudata var. *elongata* Rehder & E.H.Wilson === **Magnolia sprengeri** Pamp.

Magnolia denudata var. *emarginata* (Finet & Gagnep.) Pamp. === **Magnolia sargentiana** Rehder & E.H.Wilson

Magnolia denudata var. *fargesii* (Finet & Gagnep.) Pamp. === **Magnolia biondii** Pamp.

Magnolia denudata var. *purpurascens* (Maxim.) Rehder & E.H.Wilson === **Magnolia sprengeri** Pamp.

Magnolia denudata var. *pyramidalis* T.B.Chao & Z.X.Chen === **Magnolia denudata** Desr.

Magnolia discolor Vent. === **Magnolia liliiflora** Desr.

Magnolia diva Stapf ex Dandy === **Magnolia sprengeri** Pamp.

Magnolia × *dorsopurpurea* Makino === ?

Magnolia duclouxii (Finet & Gagnep.) Hu === **Manglietia duclouxii** Finet & Gagnep.

Magnolia duperreana Pierre === **Kmeria duperreana** (Pierre) Dandy

Magnolia echinina P.Parm. === ?

Magnolia elliptica Link. === **Magnolia grandiflora** L.

Magnolia elliptilimba Y.W.Law & Z.Y.Gao === **Magnolia zenii** W.C.Cheng

Magnolia elongata (Rehder & E.H.Wilson) Millais === **Magnolia sprengeri** Pamp.

Magnolia emarginata (Finet & Gagnep.) W.C.Cheng === **Magnolia sargentiana** Rehder & E.H.Wilson

Magnolia eriostepta Dandy ex Gagnep. === ?

Magnolia eriostepta var. *poilanei* Dandy ex Humbert === **Magnolia liliifera** (L.) Baill. var. **liliifera**

Magnolia excelsa Wall. === **Michelia doltsopa** Buch.-Ham. ex DC.

Magnolia excelsa Jacques === **Michelia doltsopa** Buch.-Ham. ex DC.

Magnolia exoniensis Millais === **Magnolia grandiflora** L.

Magnolia famasiha P.Parm. === **Magnolia salicifolia** (Siebold & Zucc.) Maxim.

Magnolia fargesii (Finet & Gagnep.) W.C.Cheng === **Magnolia biondii** Pamp.

Magnolia fasciata Vent. === **Michelia figo** (Lour.) Spreng. var. **figo**

Magnolia fasciculata P.Parm. === ?

Magnolia fatiscens Rich. ex DC. === **Magnolia dodecapetala** (Lam.) Govaerts

Magnolia ferruginea P.Parm. === ?

Magnolia ferruginea W.Watson === **Magnolia grandiflora** L.

Magnolia figo (Lour.) DC. === **Michelia figo** (Lour.) Spreng.

Magnolia fistulosa (Finet & Gagnep.) Dandy === **Magnolia championii** Benth.

Magnolia foetida (L.) Sarg. === **Magnolia grandiflora** L.

Magnolia foetida f. *margaretta* Ashe === **Magnolia grandiflora** L.

Magnolia foetida f. *parvifolia* Ashe === **Magnolia grandiflora** L.

Magnolia forbesii King === **Magnolia liliifera** (L.) Baill. var. **liliifera**

Magnolia fordiana (Oliv.) Hu === **Manglietia fordiana** Oliv.

Magnolia fragrans Salisb. === **Magnolia virginiana** L.

Magnolia fragrans Raf. === ?

Magnolia fraseri subsp. *pyramidata* (Bartram) E.Murray === **Magnolia fraseri** var. **pyramidata** (Bartram) Pamp.

Magnolia frondosa Salisb. === **Magnolia tripetala** (L.) L.

Magnolia funiushanensis T.B.Chao & al. === **Magnolia biondii** Pamp.

Magnolia funiushanensis var. *purpurea* T.B.Chao & J.T.Gao === **Magnolia biondii** Pamp.

Magnolia fuscata Andrews === **Michelia figo** (Lour.) Spreng. var. **figo**

Magnolia fuscata var. *annonifolia* (Salisb.) DC. === **Michelia figo** (Lour.) Spreng. var. **figo**

Magnolia fuscata var. *hebeclada* DC. === **Michelia figo** (Lour.) Spreng. var. **figo**

Magnolia fuscata var. *parviflora* (Blume) Steud. === **Michelia figo** (Lour.) Spreng. var. **figo**

Magnolia galissoniensis Millais === **Magnolia grandiflora** L.

Magnolia glabra P.Parm. === ?

Magnolia glauca (L.) L. === **Magnolia virginiana** L.

Magnolia glauca Thunb. === **Magnolia obovata** Thunb.

Magnolia glauca (Korth.) Pierre === **Magnolia elegans** (Blume) H.Keng

Magnolia glauca var. *argentea* DC. === **Magnolia virginiana** L.

Magnolia glauca var. *longifolia* Aiton === **Magnolia virginiana** L.

Magnolia glauca var. *major* Sims === **Magnolia × thompsoniana** de Vos
Magnolia glauca var. *thompsoniana* Loudon === **Magnolia × thompsoniana** de Vos
Magnolia globosa var. *sinensis* Rehder & E.H.Wilson === **Magnolia sieboldii** subsp. **sinensis** (Rehder & E.H.Wilson) Spongberg
Magnolia gloriosa Millais === **Magnolia grandiflora** L.
Magnolia gordoniana Steud. === **Magnolia virginiana** L.
Magnolia gracilis Salisb. === **Magnolia liliiflora** Desr.
Magnolia grandiflora var. *elliptica* W.T.Aiton === **Magnolia grandiflora** L.
Magnolia grandiflora var. *exoniensis* Loud. === **Magnolia grandiflora** L.
Magnolia grandiflora f. *galissoniensis* K.Koch === **Magnolia grandiflora** L.
Magnolia grandiflora var. *lanceolata* Aiton === **Magnolia grandiflora** L.
Magnolia grandiflora f. *lanceolata* (Aiton) Rehder === **Magnolia grandiflora** L.
Magnolia grandiflora var. *obovata* W.T.Aiton === **Magnolia grandiflora** L.
Magnolia halleana auct. === **Magnolia stellata** (Siebold & Zucc.) Maxim.
Magnolia hartwegii G.Nicholson === **Magnolia grandiflora** L.
Magnolia hartwicus G.Nicholson === **Magnolia grandiflora** L.
Magnolia heliophyla P.Parm. === ?
Magnolia heptapeta (Buc'hoz) Dandy === **Magnolia denudata** Desr.
Magnolia heptapeta f. *purpurascens* (Maxim.) H.Ohba === **Magnolia sprengeri** Pamp.
Magnolia highdownensis Dandy === **Magnolia wilsonii** (Finet & Gagnep.) Rehder
Magnolia hirsuta Thunb. === ?
Magnolia hodgsonii (Hook.f. & Thomson) H.Keng === **Magnolia liliifera** var. **obovata** (Korth.) Govaerts
Magnolia honanensis B.Y.Ding & T.B.Chao === **Magnolia biondii** Pamp.
Magnolia hondurensis A.M.Molina === **Magnolia guatemalensis** subsp. **hondurensis** (Molina) Vázquez
Magnolia honogi P.Parm. === **Magnolia obovata** Thunb.
Magnolia hookeri (Cubitt & W.W.Sm.) D.C.S.Raju & M.P.Nayar === **Manglietia hookeri** Cubitt & W.W.Sm.
Magnolia hoonokii Siebold === **Magnolia obovata** Thunb.
Magnolia × hybrida Dippel === ?
Magnolia hypoleuca Siebold & Zucc. === **Magnolia obovata** Thunb.
Magnolia hypoleuca var. *concolor* Siebold & Zucc. === **Magnolia obovata** Thunb.
Magnolia inodora DC. === ?
Magnolia insignis Blume === ?
Magnolia insignis Wall. === **Manglietia insignis** (Wall.) Blume
Magnolia insignis var. *angustifolia* (Hook.f. & Thomson) H.J.Chowdhery & P.Daniel === **Manglietia insignis** (Wall.) Blume
Magnolia insignis var. *latifolia* (Hook.f. & Thomson) H.J.Chowdhery & P.Daniel === **Manglietia insignis** (Wall.) Blume
Magnolia intermedia P.Parm. === ?
Magnolia javanica Koord. & Valeton === **Magnolia macklottii** (Korth.) Dandy var. **macklottii**
Magnolia keiskei (Makino) Ihrig === **Magnolia stellata** (Siebold & Zucc.) Maxim.
Magnolia kewensis Pearce === **Magnolia salicifolia** (Siebold & Zucc.) Maxim.
Magnolia kobus var. *borealis* Sarg. === **Magnolia kobus** DC.
Magnolia kobus var. *loebneri* (Kache) Spongberg === **Magnolia × loebneri** Kache
Magnolia kobus f. *rosea* (Veitch ex Hu) Blackburn === **Magnolia stellata** (Siebold & Zucc.) Maxim.
Magnolia kobus var. *stellata* (Siebold & Zucc.) Blackburn === **Magnolia stellata** (Siebold & Zucc.) Maxim.
Magnolia kobus f. *stellata* (Siebold & Zucc.) Maxim. === **Magnolia stellata** (Siebold & Zucc.) Maxim.
Magnolia kobushii Mayr === ?
Magnolia kunstleri King === **Magnolia liliifera** (L.) Baill. var. **liliifera**
Magnolia kwangtungensis Merr. === **Manglietia fordiana** var. **kwangtungensis** (Merr.) B.L.Chen & Noot.

Magnolia lacunosa Raf. === **Magnolia grandiflora** L.

Magnolia lanceolata Link === **Magnolia grandiflora** L.

Magnolia lanigera (Hook.f. & Thomson) H.J.Chowdhery & P.Daniel === **Magnolia villosa** (Miq.) H.Keng

Magnolia latifolia Aiton ex Dippel === **Magnolia virginiana** L.

Magnolia × *lenneana* (Lem.) Koehne === **Magnolia** × **soulangeana** Soul.-Bod.

Magnolia × *lennei* Van Houtte === **Magnolia** × **soulangeana** Soul.-Bod.

Magnolia liliifera var. *championii* (Benth.) Pamp. === **Magnolia championii** Benth.

Magnolia liliifera var. *taliensis* (W.W.Sm.) Pamp. === **Magnolia wilsonii** (Finet & Gagnep.) Rehder

Magnolia liliiflora var. *gracilis* (Salisb.) Rehder === **Magnolia liliiflora** Desr.

Magnolia liliiflora var. *nigra* (G.Nicholson) Rehder === **Magnolia liliiflora** Desr.

Magnolia linguifolia L. ex Descourt. === **Magnolia dodecapetala** (Lam.) Govaerts

Magnolia longifolia Sweet === **Magnolia grandiflora** L.

Magnolia longistyla P.Parm. === ?

Magnolia lotungensis Chun & C.H.Tsoong === **Magnolia nitida** var. **lotungensis** (Chun & C.H.Tsoong) B.L.Chen & Noot.

Magnolia macrophylla var. *ashei* (Weath.) D.L.Johnson === **Magnolia macrophylla** subsp. **ashei** (Weath.) Spongberg

Magnolia macrophylla var. *dealbata* (Zucc.) D.L.Johnson === **Magnolia macrophylla subsp. dealbata**

Magnolia major Millais === **Magnolia virginiana** L.

Magnolia mannii (King) King === **Michelia mannii** King

Magnolia martinii H.Lév. === **Michelia martinii** (H.Lév.) H.Lév.

Magnolia maxima Lodd. ex G.Don === **Magnolia grandiflora** L.

Magnolia meleagrioides DC. === **Michelia figo** (Lour.) Spreng. var. **figo**

Magnolia membranacea P.Parm. === **Michelia champaca** L. var. **champaca**

Magnolia membranacea var. *pealiana* (King) P.Parm. === **Magnolia pealiana** King

Magnolia michauxiana DC. === **Magnolia macrophylla** Michx. subsp. **macrophylla**

Magnolia michauxii Fraser ex Thouin === ?

Magnolia microphylla Ser. === **Magnolia grandiflora** L.

Magnolia mollicomata W.W.Sm. === **Magnolia campbellii** Hook.f. & Thomson

Magnolia mutabilis Regel === ?

Magnolia mutabilis (Blume) H.J.Chowdhery & P.Daniel === **Magnolia liliifera** (L.) Baill. var. **liliifera**

Magnolia nicholsoniana Rehder & E.H.Wilson === **Magnolia wilsonii** (Finet & Gagnep.) Rehder

Magnolia × *norbertiana* Dippel === **Magnolia** × **hybrida**

Magnolia nutans (Dandy) H.Keng === **Magnolia bintuluensis** (A.Agostini) Noot.

Magnolia obovata Aiton ex Link === **Magnolia grandiflora** L.

Magnolia obovata var. *denudata* (Desr.) DC. === **Magnolia denudata** Desr.

Magnolia obovata var. *soulangeana* (Soul.-Bod.) Ser. === **Magnolia** × **soulangeana** Soul.-Bod.

Magnolia obtusifolia === **Magnolia grandiflora** L.

Magnolia odoratissima Y.W.Law & R.Z.Zhou === **Magnolia championii** Benth.

Magnolia odoratissima Reinw. ex Blume === **Magnolia liliifera** (L.) Baill. var. **liliifera**

Magnolia officinalis subsp. *biloba* (Rehder & E.H.Wilson) W.C.Cheng & Y.W.Law === **Magnolia officinalis** var. **biloba** Rehder & E.H.Wilson

Magnolia officinalis var. *pubescens* C.Y.Deng === **Magnolia officinalis** Rehder & E.H.Wilson var. **officinalis**

Magnolia ovata P.Parm. === ?

Magnolia oyama Kort === **Magnolia sieboldii** K.Koch subsp. **sieboldii**

Magnolia pachyphylla Dandy === **Magnolia liliifera** (L.) Baill. var. **liliifera**

Magnolia paenetalauma Dandy === **Magnolia championii** Benth.

Magnolia parviflora Blume === **Michelia figo** (Lour.) Spreng. var. **figo**

Magnolia parviflora Siebold & Zucc. === **Magnolia sieboldii** K.Koch subsp. **sieboldii**

Magnolia parviflora var. *wilsonii* Finet & Gagnep. === **Magnolia wilsonii** (Finet & Gagnep.) Rehder

Magnolia parvifolia DC. === **Michelia figo** (Lour.) Spreng. var. **figo**

Magnolia patoricensis P.Parm. === **Magnolia portoricensis** Bello

Magnolia pensylvanica DC. === **Magnolia acuminata** (L.) L. var. **acuminata**

Magnolia persuaveolens var. *rigida* Noot. === **Magnolia persuaveolens** subsp. **rigida** Noot.

Magnolia phellocarpa (King) H.J.Chowdhery & P.Daniel === **Michelia baillonii** (Pierre) Finet & Gagnep.

Magnolia philippinensis P.Parm. === **Michelia compressa** (Maxim.) Sarg.

Magnolia pilosissima P.Parm. === **Magnolia macrophylla** Michx. subsp. **macrophylla**

Magnolia plumieri Sw. === **Magnolia dodecapetala** (Lam.) Govaerts

Magnolia praecocossima Koidz. === **Magnolia kobus** DC.

Magnolia praecox Millais === **Magnolia grandiflora** L.

Magnolia pravertiana Millais === **Magnolia grandiflora** L.

Magnolia precia Corrêa ex Vent. === **Magnolia denudata** Desr.

Magnolia proctoriana Rehder === **Magnolia salicifolia** (Siebold & Zucc.) Maxim.

Magnolia pseudokobus S.Abe & Akasawa === **Magnolia kobus** DC.

Magnolia pulneyensis P.Parm. === ?

Magnolia pumila Andrews === **Magnolia liliifera** (L.) Baill. var. **liliifera**

Magnolia pumila var. *championii* (Benth.) Finet & Gagnep. === **Magnolia championii** Benth.

Magnolia punduana Wall. === **Michelia punduana** Hook.f. & Thomson

Magnolia purpurascens (Maxim.) Makino === **Magnolia sprengeri** Pamp.

Magnolia purpurascens (Michx.) Millais === **Magnolia sprengeri** Pamp.

Magnolia purpurea Curtis === **Magnolia liliiflora** Desr.

Magnolia purpurea var. *denudata* (Desr.) Loudon === **Magnolia denudata** Desr.

Magnolia pyramidata Bartram === **Magnolia fraseri** var. **pyramidata** (Bartram) Pamp.

Magnolia quinquepeta (Buc'hoz) Dandy === **Magnolia liliiflora** Desr.

Magnolia rabaniana (Hook.f. & Thomson) D.C.S.Raju & M.P.Nayar === **Magnolia liliifera** (L.) Baill. var. **liliifera**

Magnolia roraimae Steyerm. === **Magnolia ptaritepuiana** Steyerm.

Magnolia rosea (Veitch ex Hu) Ihrig === **Magnolia stellata** (Siebold & Zucc.) Maxim.

Magnolia rotundifolia Millais === **Magnolia grandiflora** L.

Magnolia rumphii (Blume) Spreng. === **Magnolia liliifera** (L.) Baill. var. **liliifera**

Magnolia rustica DC. === **Magnolia acuminata** (L.) L. var. **acuminata**

Magnolia salicifolia var. *concolor* (Miq.) Maxim. === **Magnolia salicifolia** (Siebold & Zucc.) Maxim.

Magnolia salicifolia f. *fasciata* (Millais) Rehder === **Magnolia salicifolia** (Siebold & Zucc.) Maxim.

Magnolia salicifolia var. *fasciata* Millais === **Magnolia salicifolia** (Siebold & Zucc.) Maxim.

Magnolia sargentiana var. *robusta* Rehder & E.H.Wilson === **Magnolia sargentiana** Rehder & E.H.Wilson

Magnolia scortechinii King === **Michelia scortechinii** (King) Dandy

Magnolia selloi Spreng. === **Magnolia sellowiana** (A.St.-Hil.) Govaerts

Magnolia sericea Thunb. === ?

Magnolia shangpaensis Hu === **Manglietia insignis** (Wall.) Blume

Magnolia siamensis (Dandy) H.Keng === **Magnolia liliifera** (L.) Baill. var. **liliifera**

Magnolia simii Siebold ex Miq. === **Magnolia stellata** (Siebold & Zucc.) Maxim.

Magnolia sinensis (Rehder & E.H.Wilson) Stapf === **Magnolia sieboldii** subsp. **sinensis** (Rehder & E.H.Wilson) Spongberg

Magnolia singapurensis (Ridl.) H.Keng === **Magnolia liliifera** var. **singapurensis** (Ridl.) Govaerts

Magnolia sinica (Y.W.Law) Noot. === **Manglietia sinica** (Y.W.Law) B.L.Chen & Noot.

Magnolia sinostellata P.L.Chiu & Z.H.Chen === **Magnolia stellata** (Siebold & Zucc.) Maxim.

Magnolia slavinii Harkn. === **Magnolia salicifolia** (Siebold & Zucc.) Maxim.

Magnolia × *soulangeana* var. *brozzinii* auct. === **Magnolia** × **soulangeana** Soul.-Bod.

Magnolia × *soulangeana* var. *nigra* G.Nicholson === **Magnolia liliiflora** Desr.

Magnolia spathulata W.C.Cheng ex C.P'ei === ?

Magnolia × *speciosa* Rchb. === **Magnolia** × **hybrida** Dippel

Magnolia × *speciosa* Cels ex Dippel === **Magnolia** × **hybrida** Dippel

Magnolia spectabilis G.Nicholson === **Magnolia denudata** Desr.

Magnolia sphenocarpa Hook.f. & Thomson === **Magnolia pterocarpa** Roxb.

Magnolia splendens Reinw. ex Blume === **Magnolia liliifera** (L.) Baill. var. **liliifera**

Magnolia sprengeri var. *diva* (Stapf ex Dandy) Stapf === **Magnolia sprengeri** Pamp.

Magnolia sprengeri var. *elongata* (Rehder & E.H.Wilson) Johnstone === **Magnolia sprengeri** Pamp.

Magnolia stellata var. *keiskei* Makino === **Magnolia stellata** (Siebold & Zucc.) Maxim.

Magnolia stellata var. *rosea* Veitch ex Hu === **Magnolia stellata** (Siebold & Zucc.) Maxim.

Magnolia stricta G.Nicholson === **Magnolia grandiflora** L.

Magnolia superba G.Nicholson === **Magnolia denudata** Desr.

Magnolia talaumoides Dandy === **Magnolia championii** Benth.

Magnolia taliensis W.W.Sm. === **Magnolia wilsonii** (Finet & Gagnep.) Rehder

Magnolia tardiflora Ser. === **Magnolia grandiflora** L.

Magnolia tenuicarpella H.T.Chang === **Magnolia championii** Benth.

Magnolia thamnodes Dandy === **Magnolia liliifera** (L.) Baill. var. **liliifera**

Magnolia thurberi G.Nicholson === **Magnolia kobus** DC.

Magnolia tomentosa Ser. === **Magnolia grandiflora** L.

Magnolia tomentosa Thunb. === **Edgeworthia papyrifera** Sieb. & Zucc. (Thymelaeaceae)

Magnolia triumphans G.Nicholson === **Magnolia denudata** Desr.

Magnolia tsarongensis W.W.Sm. & Forrest === **Magnolia globosa** Hook.f. & Thomson

Magnolia umbellata Steud. === **Magnolia tripetala** (L.) L.

Magnolia umbrella Desr. === **Magnolia tripetala** (L.) L.

Magnolia velutina P.Parm. === ?

Magnolia verecunda Koidz. === **Magnolia sieboldii** K.Koch subsp. **sieboldii**

Magnolia versicolor Salisb. === **Michelia figo** (Lour.) Spreng. var. **figo**

Magnolia villariana (Rolfe) D.C.S.Raju & M.P.Nayar === **Magnolia liliifera** var. **angatensis** (Blanco) Govaerts

Magnolia virginiana var. *acuminata* L. === **Magnolia acuminata** (L.) L.

Magnolia virginiana subsp. *australis* (Sarg.) E.Murray === **Magnolia virginiana** L.

Magnolia virginiana var. *australis* Sarg. === **Magnolia virginiana** L.

Magnolia virginiana var. *foetida* L. === **Magnolia grandiflora** L.

Magnolia virginiana var. *glauca* L. === **Magnolia virginiana** L.

Magnolia virginiana var. *grisea* L. === ?

Magnolia virginiana var. *longifolia* Aiton === **Magnolia virginiana** L.

Magnolia virginiana var. *parva* Ashe === **Magnolia virginiana** L.

Magnolia virginiana var. *pumila* Nutt. === **Magnolia virginiana** L.

Magnolia virginiana var. *tripetala* L. === **Magnolia tripetala** (L.) L.

Magnolia vrieseana (Miq.) Baill. ex Pierre === **Elmerrillia ovalis** (Miq.) Dandy

Magnolia × *watsonii* Hook.f. === **Magnolia** × **wiesneri** Carr.

Magnolia wilsonii f. *nicholsoniana* (Rehder & E.H.Wilson) Rehder === **Magnolia wilsonii** (Finet & Gagnep.) Rehder

Magnolia wilsonii f. *taliensis* (W.W.Sm.) Rehder === **Magnolia wilsonii** (Finet & Gagnep.) Rehder

Magnolia xerophila P.Parm. === **Mimusops elengi** L. (Sapotaceae)

Magnolia yulan Desf. === **Magnolia denudata** Desr.

Magnolia yulan var. *soulangeana* (Soul.-Bod.) Lindl. === **Magnolia** × **soulangeana** Soul.-Bod.

Magnolia yunnanensis (Hu) Noot. === **Magnolia nitida** W.W.Sm. var. **nitida**

Manglietia

29 species, Asia and Malesia (not, however, in the Moluccas or New Guinea) (Nooteboom 1985); includes *Paramanglietia* Hu et Cheng (1951). The genus is closely related to *Magnolia*, differing in hair, leaf and gynoecial characters. Besides the papers cited here, there is incidental treatment of the genus in enthusiasts' works on magnolias. Since publication of Tiep's revision (1980) additional Asian species have been described. Those in Malesia were most recently reviewed by Nooteboom (1988; see **Malesia**) and those in China by Chen and Nooteboom (1993; see **Asia**). (Magnolieae)

- Tiep, N. V. (1980). Beiträge aur Sippenstruktur der Gattung *Manglietia* Bl. (Magnoliaceae). Feddes Rep. 91 (9-10): 497-576. Ge. — Partial revision, covering 22 species in 2 sections (key. pp. 564-565). Much attention paid to comparative vegetative anatomy, particularly of the leaf.
- Tiep, N. V., W. Vent & G. Natho (1980). Über die Gattung *Manglietia* Bl. (Magnoliaceae). Wiss. Zeitschr. Humboldt-Univ. Berlin, Math.-Naturw. Reih. 29(3): 323-328, illus., map. Ge. — Introduction to a revision of the genus; fuller treatment in Tiep (1980).
- Nooteboom, H. P. (1985). Notes on Magnoliaceae. Blumea 31(1): 65-121. En. — *Manglietia*, pp. 91-97. Includes a treatment of Malesian species (5, 2 of them new), with key.

Manglietia Blume, Verh. Batav. Genootsch. Kunsten 9: 149 (1823).
    S. China, Trop. Asia. 36 40 41 42.
        Paramanglietia Hu & W.C.Cheng, Acta Phytotax. Sin. 1: 255 (1951).
        Manglietiastrum Y.W.Law, Acta Phytotax. Sin. 17(4): 72 (1979).

Manglietia aromatica Dandy, J. Bot. 69: 231 (1931). *Paramanglietia aromatica* (Dandy) Hu & W.C.Cheng, Acta Phytotax. Sin. 1: 256 (1951).
    N. Vietnam, S. China (Yunnan, Guizhou, Guangxi). 36 CHC CHS 41 VIE. Phan.

Manglietia blaoensis Gagnep., Notul. Syst. (Paris) 8: 63 (1939). *Magnolia blaoensis* (Gagnep.) Dandy in S.Nilsson, World Pollen Spore Fl. 3(Magnoliaceae): 5 (1974).
    Vietnam. 41 VIE. Phan.

Manglietia calophylla Dandy, J. Bot. 66: 46 (1928).
    W. Sumatra. 42 SUM. Phan.

Manglietia caveana Hook.f. & Thomson, Fl. Ind. 1: 76 (1855). *Magnolia caveana* (Hook.f. & Thomson) D.C.S.Raju & M.P.Nayar, Indian J. Bot. 3: 171 (1980).
    Assam, N. Burma. 40 ASS 41 BMA. Phan.

Manglietia chevalieri Dandy, J. Bot. 68: 204 (1930).
    Laos, Vietnam. 41 LAO VIE. Phan.

Manglietia conifera Dandy, J. Bot. 68: 205 (1930).
    N. Vietnam, S. China (Yunnan, Guangdong, Guangxi). 36 CHC CHS 41 VIE. Phan.
        Manglietia chingii Dandy, J. Bot. 69: 232 (1931).
        Manglietia tenuipes Dandy, J. Bot. 69: 232 (1931).
        Manglietia glaucifolia Y.W.Law & Y.F.Wu, Guihaia 6: 263 (1986).

Manglietia dandyi (Gagnep.) Dandy in S.Nilsson, World Pollen Spore Fl. 3(Magnoliaceae): 5 (1974).
    Laos, Vietnam, China (Yunnan, Guangdong, Guangxi). 36 CHC CHS 41 LAO VIE. Phan.
      * *Magnolia dandyi* Gagnep., Notul. Syst. (Paris) 8: 63 (1939).

Manglietia dolichogyna Dandy ex Noot., Blumea 31: 95 (1985).
  Pen. Malaysia, Borneo (Sabah). 42 BOR MLY. Phan.

Manglietia duclouxii Finet & Gagnep., Bull. Soc. Bot. France 52(4): 33 (1906). *Magnolia duclouxii* (Finet & Gagnep.) Hu in H.H.Hu & W.Y.Chun, Icon. Pl. Sin. 2: 18 (1929).
  Vietnam, China (Yunnan, Sichuan, Guangxi). 36 CHC CHS 41 VIE. Phan.

Manglietia fordiana Oliv., Hooker's Icon. Pl. 20: 1953 (1891). *Magnolia fordiana* (Oliv.) Hu, J. Arnold Arbor. 5: 228 (1924).
  Vietnam, China (Anhui, Fujian, Guangdong, Guangxi, Zhejiang, Guizhou, Yunnan, Jiangxi, Hunan, Hong Kong), Hainan. 36 CHC CHH CHS 41 VIE. Phan.

  var. **calcarea** (X.H.Song) B.L.Chen & Noot., Ann. Missouri Bot. Gard. 80: 1040 (1993).
  China (Guizhou). 36 CHS. Phan.
    * *Manglietia calcarea* X.H.Song, J. Nanjing Inst. Forest. 1984(4): 46 (1984).

  var. **fordiana**
    Vietnam, China (Anhui, Fujian, Guangdong, Guangxi, Zhejiang, Guizhou, Yunnan, Jiangxi, Hunan, Hong Kong), Hainan. 36 CHC CHH CHS 41 VIE. Phan.
    Manglietia hainanensis Dandy, J. Bot. 68: 204 (1930).
    Paramanglietia microcarpa H.T.Chang, Acta Sci. Nat. Univ. Sunyatseni 1961: 55 (1961).
    Manglietia yuyuanensis Y.W.Law, Bull. Bot. Res., Harbin 5: 125 (1985).

  var. **forrestii** (W.W.Sm. ex Dandy) B.L.Chen & Noot., Ann. Missouri Bot. Gard. 80: 1040 (1993).
    China (SW. Guangxi, S. Yunnan). 36 CHC CHS. Phan.
    * *Manglietia forrestii* W.W.Sm. ex Dandy, Notes Roy. Bot. Gard. Edinburgh 16: 126 (1928).
    Manglietia globosa H.T.Chang, Acta Sci. Nat. Univ. Sunyatseni 1961: 54 (1961).

  var. **kwangtungensis** (Merr.) B.L.Chen & Noot., Ann. Missouri Bot. Gard. 80: 1041 (1993).
    China (Guangdong, Guangxi). 36 CHS. Phan.
    * *Magnolia kwangtungensis* Merr., J. Arnold Arbor. 8: 5 (1927). *Manglietia kwangtungensis* (Merr.) Dandy, Bull. Misc. Inform. Kew 1927: 264 (1927).

Manglietia garrettii Craib, Bull. Misc. Inform. Kew 1922: 166 (1922).
  Thailand, Vietnam, SW. China (S. Yunnan). 36 CHC 41 THA VIE. Phan.

Manglietia glauca Blume, Verh. Batav. Genootsch. Kunsten 9: 150 (1823).
  Sumatra, Java, Lesser Sunda Is., Sulawesi. 42 JAW LSI SUL SUM. Phan.

  var. **glauca**
    Sumatra, Java, Lesser Sunda Is., Sulawesi. 42 JAW LSI SUL SUM. Phan.
    Magnolia blumei Prantl in H.G.A.Engler & K.A.E.Prantl, Nat. Pflanzenfam. 3(2): 16 (1888).

  var. **lanuginosa** Dandy, Bull. Misc. Inform. Kew 1928: 187 (1928).
    Sumatra (around lake Tobi). 42 SUM. Phan.
    Manglietia lanuginosa (Dandy) Noot., Blumea 31: 94 (1985).

  var. **sumatrana** (Miq.) Dandy, Bull. Misc. Inform. Kew 1928: 188 (1928).
    W. Sumatra. 42 SUM. Phan.
    * *Manglietia sumatrana* Miq., Fl. Ned. Ind. Eerste Bijv.: 367 (1861).
    Manglietia pilosa P.Parm., Bull. Sc. France Belgique 27: 217, 292 (1896).
    Manglietia singalanensis A.Agostini, Atti Reale Accad. Fisiocrit. Siena, X, 1: 183 (1926).

Manglietia grandis Hu & W.C.Cheng, Acta Phytotax. Sin. 1: 158 (1951).
  SW. China (SE. Yunnan). 36 CHC. Phan.

Manglietia hebecarpa C.Y.Wu & Y.W.Law, Acta Phytotax. Sin. 34: 88 (1996).
  China (Yunnan). 36 CHC. Phan.

Manglietia hookeri Cubitt & W.W.Sm., Rec. Bot. Surv. India 4: 273 (1911). *Magnolia hookeri* (Cubitt & W.W.Sm.) D.C.S.Raju & M.P.Nayar, Indian J. Bot. 3: 171 (1980).
N. Burma, SW. China (Yunnan, Guizhou). 36 CHC 41 BMA. Phan.

Manglietia insignis (Wall.) Blume, Fl. Javae 19-20: 23 (1829).
Nepal, Assam, N. Burma, Tibet, N. Vietnam, China (Yunnan, Sichuan, Guizhou, Hunan, Hubei). 36 CHC CHS CHT 40 ASS NEP 41 BMA VIE. Phan.
  * *Magnolia insignis* Wall., Tent. Fl. Napal.: 5 (1824).
    Manglietia insignis var. *angustifolia* Hook.f. & Thomson in J.D.Hooker, Fl. Brit. Ind. 1: 42 (1872). *Magnolia insignis* var. *angustifolia* (Hook.f. & Thomson) H.J.Chowdhery & P.Daniel, Indian J. Forest. 4: 64 (1981).
    Manglietia insignis var. *latifolia* Hook.f. & Thomson in J.D.Hooker, Fl. Brit. Ind. 1: 42 (1872). *Magnolia insignis* var. *latifolia* (Hook.f. & Thomson) H.J.Chowdhery & P.Daniel, Indian J. Forest. 4: 64 (1981).
    Magnolia shangpaensis Hu, Acta Phytotax. Sin. 1: 157 (1951).
    Manglietia patungensis Hu, Acta Phytotax. Sin. 1: 335 (1951).
    Manglietia yunnanensis Hu, Acta Phytotax. Sin. 1: 159 (1951).
    Manglietia maguanica Hung T.Chang & B.L.Chen, Acta Sci. Nat. Univ. Sunyatseni 1988: 109 (1988).
    Manglietia tenuifolia Hung T.Chang & B.L.Chen, Acta Sci. Nat. Univ. Sunyatseni 1988: 110 (1988).

Manglietia lucida B.L.Chen & S.C.Yang, Acta Sci. Nat. Univ. Sunyatseni 1988(3): 94 (1988).
SW. China (SE. Yunnan). 36 CHC. Phan.

Manglietia megaphylla Hu & W.C.Cheng, Acta Phytotax. Sin. 1: 159 (1951).
China (Yunnan, Guangxi). 36 CHC CHS. Phan.

Manglietia microtricha Y.W.Law, Bull. Bot. Res., Harbin 5: 125 (1985).
Tibet (Moyuo). 36 CHT. Phan.

Manglietia moto Dandy, Notes Roy. Bot. Gard. Edinburgh 16: 128 (1928).
China (Guangdong, N. Guangxi, S. Hunan). 36 CHS. Phan.

Manglietia obovalifolia C.Y.Wu & Y.W.Law, Acta Phytotax. Sin. 34: 89 (1996).
China (Yunnan, Guizhou). 36 CHC. Phan.

Manglietia ovoidea Hung T.Chang & B.L.Chen, Acta Sci. Nat. Univ. Sunyatseni 1988: 108 (1988).
SW. China (Yunnan). 36 CHC. Phan.

Manglietia phuthoensis Dandy ex Gagnep. in P.H.Lecomte, Fl. Indo-Chine, Suppl. 1: 36 (1938).
Vietnam. 41 VIE. Phan.

Manglietia rufibarbata Dandy, Notes Roy. Bot. Gard. Edinburgh 16: 128 (1928).
Vietnam. 41 VIE. Phan.

Manglietia sabahensis Dandy ex Noot., Blumea 31: 95 (1985).
Borneo (Sabah: Mt. Kinabalu). 42 BOR. Phan.

Manglietia sinica (Y.W.Law) B.L.Chen & Noot., Ann. Missouri Bot. Gard. 80: 1051 (1993).
China (SE. Yunnan). 36 CHC. Phan.
  * *Manglietiastrum sinicum* Y.W.Law, Acta Phytotax. Sin. 17(4): 73 (1979). *Magnolia sinica* (Y.W.Law) Noot., Blumea 31: 91 (1985).

Manglietia szechuanica Hu, Bull. Fan Mem. Inst. Biol. 10: 117 (1940).
  SW. China (N. Yunnan, S. & C. Sichuan). 36 CHC. Phan.

Manglietia utilis Dandy, Bull. Misc. Inform. Kew 1927: 310 (1927).
  N. Burma. 41 BMA.

Manglietia ventii N.V.Tiep, Repert. Spec. Nov. Regni Veg. 91: 560 (1980).
  SW. China (SE. Yunnan). 36 CHC. Phan.

Synonyms:
Manglietia calcarea X.H.Song === **Manglietia fordiana** var. **calcarea** (X.H.Song) B.L.Chen & Noot.
Manglietia celebica Miq. === **Magnolia liliifera** (L.) Baill. var. **liliifera**
Manglietia chingii Dandy === **Manglietia conifera** Dandy
Manglietia crassipes Y.W.Law === **Manglietia pachyphylla**
Manglietia forrestii W.W.Sm. ex Dandy === **Manglietia fordiana** var. **forrestii** (W.W.Sm. ex Dandy) B.L.Chen & Noot.
Manglietia glaucifolia Y.W.Law & Y.F.Wu === **Manglietia conifera** Dandy
Manglietia globosa H.T.Chang === **Manglietia fordiana** var. **forrestii** (W.W.Sm. ex Dandy) B.L.Chen & Noot.
Manglietia hainanensis Dandy === **Manglietia fordiana** Oliv. var. **fordiana**
Manglietia insignis var. *angustifolia* Hook.f. & Thomson === **Manglietia insignis** (Wall.) Blume
Manglietia insignis var. *latifolia* Hook.f. & Thomson === **Manglietia insignis** (Wall.) Blume
Manglietia kwangtungensis (Merr.) Dandy === **Manglietia fordiana** var. **kwangtungensis** (Merr.) B.L.Chen & Noot.
Manglietia lanuginosa (Dandy) Noot. === **Manglietia glauca** var. **lanuginosa** Dandy
Manglietia macklottii Korth. === **Magnolia macklottii** (Korth.) Dandy
Manglietia maguanica Hung T.Chang & B.L.Chen === **Manglietia insignis** (Wall.) Blume
Manglietia minahassae Koord. & Valeton ex Koord. === **Madhuca burckiana**
Manglietia oortii Korth. === **Magnolia elegans** (Blume) H.Keng
Manglietia pachyphylla H.T.Chang === ?
Manglietia patungensis Hu === **Manglietia insignis** (Wall.) Blume
Manglietia pilosa P.Parm. === **Manglietia glauca** var. **sumatrana** (Miq.) Dandy
Manglietia scortechinii King === **Michelia scortechinii** (King) Dandy
Manglietia sebassa King === **Magnolia liliifera** (L.) Baill. var. **liliifera**
Manglietia singalanensis A.Agostini === **Manglietia glauca** var. **sumatrana** (Miq.) Dandy
Manglietia sumatrana Miq. === **Manglietia glauca** var. **sumatrana** (Miq.) Dandy
Manglietia tenuifolia Hung T.Chang & B.L.Chen === **Manglietia insignis** (Wall.) Blume
Manglietia tenuipes Dandy === **Manglietia conifera** Dandy
Manglietia thamnodes (Dandy) Gagnep. === **Magnolia liliifera** (L.) Baill. var. **liliifera**
Manglietia wangii Hu === **Magnolia henryi** Dunn
Manglietia yunnanensis Hu === **Manglietia insignis** (Wall.) Blume
Manglietia yuyuanensis Y.W.Law === **Manglietia fordiana** Oliv. var. **fordiana**

Manglietiastrum

Synonyms:
Manglietiastrum Y.W.Law === **Manglietia** Blume
Manglietiastrum sinicum Y.W.Law === **Manglietia sinica** (Y.W.Law) B.L.Chen & Noot.

Michelia

47 species, S and SE Asia to Japan and into Malesia (exclusive of Sulawesi and New Guinea); includes *Tsoongiodendron* Chun (1963) and *Paramichelia* Hu (1940). No recent overall treatment is available. Chinese species are now covered in Chen and Nooteboom (1993; see **Asia**) and those from Malesia also in Nooteboom (1988; see **Malesia**). Limited treatment is also available in enthusiasts' works on magnolias. (Michelieae)

- Nooteboom, H. P. (1985). Notes on Magnoliaceae. Blumea 31(1): 65-121. En. — *Michelia*, pp. 108-121; regional revision (8 species), with key. A very full synonymy for *M. champaca*, long cultivated, is included.

Michelia L., Sp. Pl.: 536 (1753).
   Trop. & Subtrop. Asia. 00 36 38 40 41 42.
      Champaca Adans., Fam. Pl. 2: 365 (1763).
      Liriopsis Spach, Hist. Nat. Vég. 7: 460 (1839).
      Sampacca Kuntze, Revis. Gen. Pl.: 6 (1891).
      Paramichelia Hu, Sunyatsenia 4: 142 (1940).
      Tsoongiodendron Chun, Acta Phytotax. Sin. 8: 281 (1963).

Michelia aenea Dandy, J. Bot. 68: 211 (1930).
   Vietnam, SW. China (SE. Yunnan). 36 CHC 41 VIE. Phan.
      Michelia oblongifolia Hung T.Chang & B.L.Chen, Acta Sci. Nat. Univ. Sunyatseni 1987(3): 86 (1987).
      Michelia longistyla Y.W.Law & Y.F.Wu, Acta Bot. Yunnan. 10: 341 (1988).

Michelia × alba DC., Syst. Nat. 1: 449 (1817). M. champaca × M. montana.
   Cult. 00 CUL. Phan.
      Michelia × longifolia Blume, Verh. Batav. Genootsch. Kunsten 9: 155 (1823). *Sampacca × longifolia* (Blume) Kuntze, Revis. Gen. Pl.: 6 (1891).
      Michelia × longifolia var. *racemosa* Blume, Fl. Javae 19-20: 13 (1829).

Michelia angustioblonga Y.W.Law & Y.F.Wu, Bull. Bot. Res., Harbin 6: 97 (1986).
   SW. China (Guizhou). 36 CHC. Nanophan. or phan.

Michelia baillonii (Pierre) Finet & Gagnep., Bull. Soc. Bot. France 52(4): 46 (1906).
   Assam, Burma, Thailand, Cambodia, Vietnam, SW. China (S. Yunnan). 36 CHC 40 ASS 41 BMA CBD THA VIE. Phan.
      Magnolia baillonii Pierre, Fl. Forest. Cochinch.: 2 (1880). *Aromadendron baillonii* (Pierre) Craib, Fl. Siam.: 26 (1925). *Paramichelia baillonii* (Pierre) Hu, Sunyatsenia 4: 144 (1940).
      Talauma phellocarpa King, Ann. Roy. Bot. Gard. (Calcutta) 3(2): 205 (1891). *Michelia phellocarpa* (King) Finet & Gagnep., Bull. Soc. Bot. France 52(4): 44 (1906). *Magnolia phellocarpa* (King) H.J.Chowdhery & P.Daniel, Indian J. Forest. 4: 64 (1981).
      Talauma spongocarpa King, Ann. Roy. Bot. Gard. (Calcutta) 3(2): 205 (1891).
      Aromadendron spongocarpum (King) Craib, Fl. Siam.: 25 (1925).

Michelia balansae (A.DC.) Dandy, Bull. Misc. Inform. Kew 1927: 263 (1927).
   Vietnam, Hainan, China (Yunnan, Guangxi, Fujiang). 36 CHC CHH CHS 41 VIE. Phan.
      Magnolia balansae A.DC., Bull. Herb. Boissier, II, 4: 294 (1904).
      Michelia baviensis Finet & Gagnep., Bull. Soc. Bot. France 52(4): 44 (1906).
      Michelia tonkinensis A.Chev., Bull. Econ. Indochine 20: 792 (1918).
      Michelia balansae var. *appressipubescens* Y.W.Law, Bull. Bot. Res., Harbin 5: 124 (1985).
      Michelia balansae var. *brevipes* B.L.Chen, Acta Sci. Nat. Univ. Sunyatseni 1988: 112 (1988).

Michelia braianensis Gagnep., Notul. Syst. (Paris) 8: 64 (1939). *Paramichelia braianensis* (Gagnep.) Dandy in S.Nilsson, World Pollen Spore Fl. 3(Magnoliaceae): 5 (1974). Vietnam. 41 VIE. Phan.

Michelia cavaleriei Finet & Gagnep., Bull. Soc. Bot. France 53(4): 573 (1906).
China (Yunnan, Sichuan, Hubei, Hunan, Guizhou, Guangxi, Guangdong, Fujiang). 36 CHC CHS. Phan.
Michelia platypetala Hand.-Mazz., Anz. Kaiserl. Akad. Wiss. Wien, Math.-Naturwiss. Kl. 58: 89 (1921).
Michelia fallax Dandy, Notes Roy. Bot. Gard. Edinburgh 16: 130 (1928).
Michelia elegans Y.W.Law & Y.F.Wu, Bull. Bot. Res., Harbin 8(3): 71 (1988).

Michelia champaca L., Sp. Pl.: 536 (1753). *Magnolia champaca* (L.) Baill. ex Pierre, Fl. Forest. Cochinch.: 3 (1880).
S. India to Lesser Sunda Is. 36 CHC CHT 40 IND 41 BMA VIE 42 JAW LSI MLY SUM. Phan.

var. **champaca**
S. India, Burma, Vietnam, SE. Tibet, SW. China (S. Yunnan). 36 CHC CHT 40 IND 41 BMA VIE. Phan.
Michelia euonymoides Burm.f., Fl. Indica: 124 (1768).
Champaca michelia Noronha, Verh. Batav. Genootsch. Kunsten, 5(4): 11 (1790).
Michelia sericea Pers., Syn. Pl. 2: 94 (1806).
Michelia suaveolens Pers., Syn. Pl. 2: 94 (1806). *Sampacca suaveolens* (Pers.) Kuntze, Revis. Gen. Pl.: 6 (1891).
Michelia rufinervis DC., Syst. Nat. 1: 449 (1817).
Michelia aurantiaca Wall., Pl. Asiat. Rar. 2: 39 (1831).
Michelia rheedei Wight, Ill. Ind. Bot. 1: 14 (1840).
Michelia blumei Steud., Nomencl. Bot., ed. 2, 2: 139 (1841).
Michelia champaca var. *blumei* Moritzi in H.Zollinger, Syst. Verz.: 36 (1846).
Michelia tsiampacca var. *blumei* Moritzi in H.Zollinger, Syst. Verz.: 36 (1846), orth. var.
Michelia champava Lour. ex Gomes, Mem. Acad. Sc. Lisboa Cl. Sc. Pol. Mor. Bel.-Let. n.s. 4: 27 (1868).
Talauma villosa f. *celebica* Miq., Ann. Mus. Bot. Lugduno-Batavi 4: 70 (1868).
Sampacca euonymoides Kuntze, Revis. Gen. Pl.: 6 (1891).
Sampacca velutina Kuntze, Revis. Gen. Pl.: 6 (1891).
Magnolia membranacea P.Parm., Bull. Sc. France Belgique 27: 200, 258 (1896).

var. **pubinervia** (Blume) Miq., Ann. Mus. Bot. Lugduno-Batavi 4: 72 (1868).
Pen. Malaysia, Sumatra, Java, Lesser Sunda Is. (Sumbawa). 42 JAW LSI MLY SUM. Phan.
*Michelia pubinervia Blume, Fl. Javae 19-20: 14 (1829).
Michelia velutina Blume, Fl. Javae 19-20: 17 (1829). *Champaca velutina* Kuntze, Revis. Gen. Pl.: 6 (1891).
Talauma villosa f. *celebica* Miq., Ann. Mus. Bot. Lugduno-Batavi 4: 70 (1868).
Michelia pilifera Bakh.f., Blumea 12: 61 (1963).

Michelia chapensis Dandy, J. Bot. 67: 223 (1929).
N. Vietnam, China (Yunnan, Guizhou, Guangdong, Guangxi, Hunan, Jiangxi). 36 CHC CHS 41 VIE. Phan.
Michelia constricta Dandy, J. Bot. 67: 223 (1929).
Michelia tsoi Dandy, J. Bot. 68: 213 (1930).
Michelia glaberrima H.T.Chang, Acta Sci. Nat. Univ. Sunyatseni 1961: 56 (1961).
Michelia brachyandra B.L.Chen & S.C.Yang, Acta Sci. Nat. Univ. Sunyatseni 1988(3): 98 (1988).
Michelia chartacea B.L.Chen & S.C.Yang, Acta Sci. Nat. Univ. Sunyatseni 1988(3): 97 (1988).
Michelia microcarpa B.L.Chen & S.C.Yang, Acta Sci. Nat. Univ. Sunyatseni 1988(3): 96 (1988).
Michelia nitida B.L.Chen, Acta Sci. Nat. Univ. Sunyatseni 1988: 111 (1988).

Michelia compressa (Maxim.) Sarg., Forest Fl. Japan: 11 (1894).

S. Japan, Nansei-shoto, Taiwan, SW. China (E. & S. Yunnan), Philippines. 36 CHC 38 JAP NNS TAI 42 PHI. Phan.

* *Magnolia compressa* Maxim., Bull. Acad. Imp. Sc. Saint-Pétersbourg 17: 417 (1872).

Magnolia philippinensis P.Parm., Bull. Sc. France Belgique 27: 206, 270 (1896). *Michelia philippinensis* (P.Parm.) Dandy, Bull. Misc. Inform. Kew 1927: 263 (1927).

Michelia parviflora Merr., Publ. Bur. Sci. Gov. Lab. 35: 70 (1906).

Michelia cumingii Merr. & Rolfe, Philipp. J. Sci. 3: 100 (1908).

Michelia compressa var. *formosana* Kaneh., Trans. Nat. Hist. Soc. Taiwan 20: 384 (1930). *Michelia formosana* (Kaneh.) Masam. & Suzuki, Rep. (Annual) Taihoku Bot. Gard. 3: 57 (1933).

Michelia compressa var. *macrantha* Hatus., Fl. Ryukyus: 285 (1971).

Michelia iteophylla C.Y.Wu ex Y.W.Law & Y.F.Wu, Acta Bot. Yunnan. 10: 337 (1988).

Michelia coriacea Hung T.Chang & B.L.Chen, Acta Sci. Nat. Univ. Sunyatseni 1987(3): 89 (1987).

SW. China (SE. Yunnan). 36 CHC. Phan.

Michelia polyneura C.Y.Wu ex Y.W.Law & Y.F.Wu, Acta Bot. Yunnan. 10: 340 (1988).

Michelia doltsopa Buch.-Ham. ex DC., Syst. Nat. 1: 448 (1817).

Nepal, Bhutan, Assam, N. Burma, Tibet, SW. China (Yunnan). 36 CHC CHT 40 ASS BHU NAP 41 BMA. Phan.

Magnolia excelsa Wall., Tent. Fl. Nepal.: 5 (1824). *Michelia excelsa* (Wall.) Blume, Fl. Javae 19-20: 9 (1829). *Sampacca excelsa* (Wall.) Kuntze, Revis. Gen. Pl.: 6 (1891).

Magnolia excelsa Jacques, J. Soc. Imp. Centr. Hort. 3: 476 (1857).

Michelia calcuttensis P.Parm., Bull. Sc. France Belgique 27: 214, 283 (1896).

Michelia manipurensis Watt ex Brandis, Indian Trees: 8 (1906).

Michelia wardii Dandy, Bull. Misc. Inform. Kew 1929: 222 (1929).

Michelia opipara Hung T.Chang & B.L.Chen, Acta Sci. Nat. Univ. Sunyatseni 1987(3): 90 (1987).

Michelia elliptilimba B.L.Chen & Noot., Ann. Missouri Bot. Gard. 80: 1064 (1993).

China (Yunnan). 36 CHC. Phan.

Michelia figo (Lour.) Spreng., Syst. Veg. 2: 643 (1825).

SE. China (Anhui, Guangdong, Guangxi, Jiangxi, Zhejiang). 36 CHS. Nanophan. or phan.

* *Liriodendron figo* Lour., Fl. Cochinch.: 347 (1790). *Magnolia figo* (Lour.) DC., Syst. Nat. 1: 460 (1817).

var. **crassipes** (Y.W.Law) B.L.Chen & Noot., Ann. Missouri Bot. Gard. 80: 1085 (1993).

China (N. Guangdong, NE. Guangxi). 36 CHS. Nanophan. or phan.

* *Michelia crassipes* Y.W.Law, Bull. Bot. Res., Harbin 5: 121 (1985).

var. **figo**

SE. China (Anhui, Guangdong, Guangxi, Jiangxi, Zhejiang). 36 CHS. Nanophan. or phan.

Magnolia fuscata Andrews, Bot. Repos.: 229 (1802). *Michelia fascicata* (Andrews) Vent., Jard. Malmaison: 24 (1803). *Michelia fuscata* (Andrews) Blume, Fl. Javae 19-20: 8 (1829). *Liriopsis fuscata* (Andrews) Spach, Hist. Nat. Vég. 7: 461 (1839).

Magnolia fasciata Vent., Jard. Malmaison: 24 (1803).

Magnolia annonifolia Salisb., Parad. Lond.: 5 (1806). *Magnolia fuscata* var. *annonifolia* (Salisb.) DC., Syst. Nat. 1: 458 (1817).

Magnolia versicolor Salisb., Parad. Lond.: 5 (1806).

Magnolia fuscata var. *hebeclada* DC., Syst. Nat. 1: 458 (1817).

Magnolia meleagrioides DC., Syst. Nat. 1: 458 (1817).

Magnolia parvifolia DC., Syst. Nat. 1: 459 (1817). *Michelia parvifolia* (DC.) B.D.Jacks., Index Kew. 2: 223 (1894).

Michelia parviflora Deless., Icon. Sel. Pl. 1: 22 (1821). *Sampacca parviflora* (Deless.) Kuntze, Revis. Gen. Pl.: 6 (1891).

Magnolia parviflora Blume, Bijdr.: 9 (1825). *Magnolia fuscata* var. *parviflora* (Blume) Steud., Nomencl. Bot., ed. 2, 2: 89 (1841).
Michelia skinneriana Dunn, J. Linn. Soc., Bot. 38: 354 (1908).
Michelia amoena Q.F.Zheng & M.M.Lin, Bull. Bot. Res., Harbin 7: 63 (1987).
Michelia brevipes Y.K.Li & X.M.Wang, Acta Phytotax. Sin. 25: 408 (1987).

Michelia flaviflora Y.W.Law & Y.F.Wu, Acta Bot. Yunnan. 10: 340 (1988).
Vietnam, SW. China (Yunnan). 36 CHC 41 VIE. Phan.

Michelia floribunda Finet & Gagnep., Bull. Soc. Bot. France 52(4): 46 (1906).
Thailand, Laos, Vietnam, China (Yunnan, Sichuan, Guizhou, Guangxi, Jiangxi). 36 CHC CHS 41 LAO THA VIE. Phan.
Michelia kerrii Craib, Bull. Misc. Inform. Kew 1922: 166 (1922).

Michelia foveolata Merr. ex Dandy, J. Bot. 66: 360 (1928).
Vietnam, Hainan, SW. & S. China. 36 CHC CHH CHS 41 VIE. Phan.
Michelia fulgens Dandy, J. Bot. 68: 210 (1930).
Michelia foveolata var. *cinerascens* Y.W.Law & Y.F.Wu, Bull. Bot. Res., Harbin 6: 99 (1986).

Michelia fujianensis Q.F.Zheng, Bull. Bot. Res., Harbin 1: 92 (1981).
SE. China (Fujian). 36 CHS. Phan.

Michelia fulva Hung T.Chang & B.L.Chen, ActaSci. Nat. Univ. Sunyatseni 1987(3): 87 (1987).
SW. China (Yunnan). 36 CHC. Phan.

Michelia hypolampra Dandy, J. Bot. 66: 321 (1928).
Vietnam, Hainan, China (S. Yunnan, Guangxi). 36 CHC CHH CHS 41 VIE. Phan.
Talauma gioii A.Chev., Bull. Econ. Indochine 20: 790 (1918).
Michelia hedyosperma Y.W.Law, Bull. Bot. Res., Harbin 5: 123 (1985).

Michelia ingrata B.L.Chen & S.C.Yang, Acta Sci. Nat. Univ. Sunyatseni 1988(3): 95 (1988).
SW. China (E. & S. Yunnan). 36 CHC. Phan.
Michelia calcicola C.Y.Wu ex Y.W.Law & Y.F.Wu, Acta Bot. Yunnan. 10: 339 (1988).

Michelia jiangxiensis Hung T.Chang & B.L.Chen, Acta Sci. Nat. Univ. Sunyatseni 31: 77 (1992).
S. China (Jiangxi). 36 CHC. Phan.

Michelia kingii Dandy, J. Bot. 66: 321 (1928).
Sikkim, Assam, Bangladesh. 40 ASS BAN BHU. Phan.

Michelia kisopa Buch.-Ham. ex DC., Syst. Nat. 1: 448 (1817). *Sampacca kisopa* (Buch.-Ham. ex DC.) Kuntze, Revis. Gen. Pl.: 6 (1891).
Nepal, Bhutan, Sikkim, NE. India, Tibet, Vietnam. 36 CHT 40 BHU IND NEP. Phan.
Michelia zila Buch.-Ham. ex Madden, Trans. Bot. Soc. Edinburgh 5: 127 (1858).

Michelia koordersiana Noot., Blumea 31: 111 (1985).
Pen. Malaysia, W. Sumatra. 42 MLY SUM. Phan.

Michelia lacei W.W.Sm., Notes Roy. Bot. Gard. Edinburgh 12: 216 (1920).
Burma, Vietnam, SW. China (S. Yunnan). 36 CHC 41 BMA VIE. Phan.
Michelia uniflora Dandy, Bull. Misc. Inform. Kew 1927: 203 (1927).
Michelia tignifera Dandy, J. Bot. 68: 213 (1930).
Michelia magnifica Hu, Bull. Fan Mem. Inst. Biol. 10: 118 (1940).
Michelia pachycarpa Y.W.Law & R.Z.Zhou, Bull. Bot. Res., Harbin 7: 85 (1987).

Michelia leveilleana Dandy, Bull. Misc. Inform. Kew 1927: 263 (1927).
   SW. China (Yunnan, E. Sichuan, W. Hubei, Guizhou). 36 CHC. Phan.
      Michelia cavaleriei H.Lév., Repert. Spec. Nov. Regni Veg. 9: 459 (1911).
      Michelia chongjiangensis Y.K.Li & X.M.Wang, Guizhou Sin. 3: 18 (1983).
      Michelia longipetiolata C.Y.Wu ex Y.W.Law & Y.F.Wu, Acta Bot. Yunnan. 10: 336 (1988).

Michelia macclurei Dandy, J. Bot. 66: 360 (1928).
   N. Vietnan, Hainan, China (Yunnan, Guangdong, Guangxi). 36 CHC CHH CHS 41 VIE.
      Phan.
   Michelia macclurei var. *sublanea* Dandy, J. Bot. 68: 212 (1930).

Michelia mannii King, Ann. Roy. Bot. Gard. (Calcutta) 3(2): 218 (1891). *Magnolia mannii*
   (King) King, Ann. Roy. Bot. Gard. (Calcutta) 3(2): 218 (1891).
   Assam. 40 ASS. Phan.

Michelia martinii (H.Lév.) H.Lév., Fl. Kouy-Tchéou: 270 (1915).
   Vietnam, China (Yunnan, Sichuan, Guangdong, Guizhou, Guangxi). 36 CHC CHS 41
      VIE. Phan.
      * Magnolia martinii H.Lév., Bull. Soc. Agric. Sarthe: 6 (1904).
      Michelia bodinieri Finet & Gagnep., Bull. Soc. Bot. France 53(4): 574 (1906).
      Michelia longistamina Y.W.Law, Bull. Bot. Res., Harbin 5: 122 (1985).

Michelia masticata Dandy, J. Bot. 67: 222 (1929).
   Laos, Vietnam, SW. China (Yunnan). 36 CHC 41 LAO VIE. Phan.
      Michelia sphaerantha C.Y.Wu ex Y.W.Law & Y.F.Wu, Acta Bot. Yunnan. 10: 335 (1988).

Michelia maudiae Dunn, J. Linn. Soc., Bot. 38: 353 (1908).
   Hainan, China (Guangdong, Guangxi, Guizhou, Jiangxi, Zhejiang). 36 CHC CHH CHS.
      Phan.
      Michelia chingii W.C.Cheng, Contrib. Biol. Lab. Chin. Assoc. Advancem. Sci., Sect. Bot.
         10: 110 (1936)

Michelia mediocris Dandy, J. Bot. 66: 47 (1928).
   Cambodia, Vietnam, Hainan, China (Guangdong, Guangxi, Guizhou). 36 CHC CHH CHS
      41 CBD VIE. Phan.
      Michelia mediocris var. *angustifolia* G.A.Fu, Guihaia 12: 5 (1992).

Michelia microtricha Hand.-Mazz., Anz. Kaiserl. Akad. Wiss. Wien, Math.-Naturwiss. Kl.
   58: 145 (1921).
   SW. China (Yunnan). 36 CHC. Phan.

Michelia montana Blume, Verh. Batav. Genootsch. Kunsten 9: 153 (1823). *Sampacca*
   montana (Blume) Kuntze, Revis. Gen. Pl.: 6 (1891).
   Pen. Malaysia, Sumatra (incl. Bangka), Java, Lesser Sunda Is. (Bali), Borneo (Sabah, E.
      Kalimantan). 42 BOR JAW LSI MLY SUM. Phan.
      Michelia ecicatrisata Miq., Fl. Ned. Ind. Eerste Bijv.: 368 (1861).
      Michelia montana var. *subvelutina* Miq., Ann. Mus. Bot. Lugduno-Batavi 4: 73 (1868).

Michelia nilagirica Zenker, Pl. Ind.: 21 (1835). *Sampacca nilagirica* (Zenker) Kuntze, Revis.
   Gen. Pl.: 6 (1891).
   W. & S. India, Sri Lanka. 40 IND SRI. Phan.
      Michelia glauca Wight, Ill. Ind. Bot. 1: 14 (1840).
      Michelia ovalifolia Wight, Ill. Ind. Bot. 1: 13 (1840). *Michelia nilagirica* var. *wightii*
         Hook.f. & Thomson in J.D.Hooker, Fl. Brit. Ind. 1: 44 (1872).
      Michelia pulneyensis Wight, Ill. Ind. Bot. 1: 14 (1840).
      Michelia walkeri Wight, Ill. Ind. Bot. 1: 13 (1840). *Michelia nilagirica* var. *walkeri* (Wight)
         Hook.f. & Thomson in J.D.Hooker, Fl. Brit. Ind. 1: 44 (1872).

Michelia oblonga Wall. ex Hook.f. & Thomson, Fl. Ind. 1: 81 (1855). *Sampacca oblonga* (Wall. ex Hook.f. & Thomson) Kuntze, Revis. Gen. Pl.: 6 (1891).
Assam (Khasia). 40 ASS. Phan.
Michelia lactea Buch.-Ham. ex Wall., Numer. List: 6491 (1832), nom. inval.

Michelia odora (Chun) Noot. & B.L.Chen, Ann. Missouri Bot. Gard. 80: 1086 (1993).
N. Vietnam, China (Guangdong, Guangxi, Fujian, Hainan). 36 CHH CHS 41 VIE. Phan.
* *Tsoongiodendron odorum* Chun, Acta Phytotax. Sin. 8: 283 (1963).

Michelia punduana Hook.f. & Thomson, Fl. Ind. 1: 81 (1855). *Sampacca punduana* (Hook.f. & Thomson) Kuntze, Revis. Gen. Pl.: 6 (1891).
Assam (Khasia). 40 ASS. Phan.
Magnolia punduana Wall., Numer. List: 974 (1829), nom. nud.

Michelia rajaniana Craib, Bull. Misc. Inform. Kew 1922: 225 (1922).
Thailand. 41 THA. Phan.

Michelia salicifolia A.Agostini, Atti Reale Accad. Fisiocrit. Siena, IX, 7: 23 (1926).
W. Sumatra. 42 SUM. Phan.
Michelia sumatrae Dandy, Bull. Misc. Inform. Kew 1928: 188 (1928).

Michelia scortechinii (King) Dandy, Bull. Misc. Inform. Kew 1927: 262 (1927).
Pen. Malaysia, W. Sumatra. 42 MLY SUM. Phan.
* *Manglietia scortechinii* King, J. Asiat. Soc. Bengal 58(2): 370 (1889). *Paramichelia scortechinii* (King) Dandy in S.Nilsson, World Pollen Spore Fl. 3(Magnoliaceae): 5 (1974).
Magnolia scortechinii King, Ann. Roy. Bot. Gard. (Calcutta) 3(2): 213 (1891).

Michelia shiluensis Chun & Y.F.Wu, Acta Phytotax. Sin. 8: 286 (1963).
Hainan. 36 CHH. Phan.

Michelia subulifera Dandy, J. Bot. 68: 212 (1930).
Vietnam. 41 VIE. Phan.

Michelia velutina DC., Prodr. 2: 79 (1824).
Nepal, Bhutan, Assam, Tibet, SW. China (Yunnan). 36 CHC CHT 40 ASS BHU NEP. Phan.
Michelia lanuginosa Wall., Tent. Fl. Nepal.: 8 (1824). *Sampacca lanuginosa* (Wall.) Kuntze, Revis. Gen. Pl.: 6 (1891).
Michelia lanceolata E.H.Wilson, J. Arnold Arbor. 7: 237 (1926).

Michelia wilsonii Finet & Gagnep., Bull. Soc. Bot. France 52(4): 45 (1906).
China (Sichuan, Guizhou, Jiangxi). 36 CHC CHS. Phan.
Michelia sinensis Hemsl. & E.H.Wilson, Bull. Misc. Inform. Kew 1906: 149 (1906).
Michelia szechuanica Dandy, Notes Roy. Bot. Gard. Edinburgh 16: 131 (1928).

Michelia xanthantha C.Y.Wu ex Y.W.Law & Y.F.Wu, Acta Bot. Yunnan. 10: 338 (1988).
SW. China (Yunnan: Xishiangbanna). 36 CHC. Phan.

Michelia yunnanensis Franch. ex Finet & Gagnep., Bull. Soc. Bot. France 52(4): 43 (1906).
SW. China (Yunnan, Sichuan, Guizhou). 36 CHC. Nanophan.
Michelia yunnanensis var. *angustifolia* Finet & Gagnep., Bull. Soc. Bot. France 52(4): 44 (1905).
Michelia dandyi Hu, Bull. Fan Mem. Inst. Biol. 8: 34 (1937).
Michelia laevifolia Y.W.Law & Y.F.Wu, Bull. Bot. Res., Harbin 8(3): 72 (1988).

Synonyms:

Michelia acris Ruiz & Pav. ex Lopez === **Drimys winteri** J. R. Forst. & G. Forst. (Winteraceae)

Michelia amoena Q.F.Zheng & M.M.Lin === **Michelia figo** (Lour.) Spreng. var. **figo**

Michelia arfakiana A.Agostini === **Elmerrillia tsiampacca** (L.) Dandy subsp. **tsiampacca**

Michelia aurantiaca Wall. === **Michelia champaca** L. var. **champaca**

Michelia balansae var. *appressipubescens* Y.W.Law === **Michelia balansae** (A.DC.) Dandy

Michelia balansae var. *brevipes* B.L.Chen === **Michelia balansae** (A.DC.) Dandy

Michelia baviensis Finet & Gagnep. === **Michelia balansae** (A.DC.) Dandy

Michelia beccariana A.Agostini === **Magnolia macklottii** var. **beccariana** (A.Agostini) Noot.

Michelia blumei Steud. === **Michelia champaca** L. var. **champaca**

Michelia bodinieri Finet & Gagnep. === **Michelia martinii** (H.Lév.) H.Lév.

Michelia brachyandra B.L.Chen & S.C.Yang === **Michelia chapensis** Dandy

Michelia brevipes Y.K.Li & X.M.Wang === **Michelia figo** (Lour.) Spreng. var. **figo**

Michelia caerulea DC. === ?

Michelia calcicola C.Y.Wu ex Y.W.Law & Y.F.Wu === **Michelia ingrata** B.L.Chen & S.C.Yang

Michelia calcuttensis P.Parm. === **Michelia doltsopa** Buch.-Ham. ex DC.

Michelia caloptila Y.W.Law & Y.F.Wu === ?

Michelia cathcartii Hook.f. & Thomson === **Magnolia cathcartii** (Hook.f. & Thomson) Noot.

Michelia cavaleriei H.Lév. === **Michelia leveilleana** Dandy

Michelia celebica Koord. === **Elmerrillia tsiampacca** (L.) Dandy subsp. **tsiampacca**

Michelia champaca var. *blumei* Moritzi === **Michelia champaca** L. var. **champaca**

Michelia champava Lour. ex Gomes === **Michelia champaca** L. var. **champaca**

Michelia chartacea B.L.Chen & S.C.Yang === **Michelia chapensis** Dandy

Michelia chingii W.C.Cheng === **Michelia maudiae** Dunn

Michelia chongjiangensis Y.K.Li & X.M.Wang === **Michelia leveilleana** Dandy

Michelia coerulea Steud. === **Michelia caerulea**

Michelia compressa var. *formosana* Kaneh. === **Michelia compressa** (Maxim.) Sarg.

Michelia compressa var. *macrantha* Hatus. === **Michelia compressa** (Maxim.) Sarg.

Michelia constricta Dandy === **Michelia chapensis** Dandy

Michelia crassipes Y.W.Law === **Michelia figo** var. **crassipes** (Y.W.Law) B.L.Chen & Noot.

Michelia cumingii Merr. & Rolfe === **Michelia compressa** (Maxim.) Sarg.

Michelia dandyi Hu === **Michelia yunnanensis** Franch. ex Finet & Gagnep.

Michelia ecicatrisata Miq. === **Michelia montana** Blume

Michelia elegans Y.W.Law & Y.F.Wu === **Michelia cavaleriei** Finet & Gagnep.

Michelia euonymoides Burm.f. === **Michelia champaca** L. var. **champaca**

Michelia excelsa (Wall.) Blume === **Michelia doltsopa** Buch.-Ham. ex DC.

Michelia fallax Dandy === **Michelia cavaleriei** Finet & Gagnep.

Michelia fascicata (Andrews) Vent. === **Michelia figo** (Lour.) Spreng. var. **figo**

Michelia forbesii Baker f. === **Elmerrillia tsiampacca** (L.) Dandy subsp. **tsiampacca**

Michelia formosana (Kaneh.) Masam. & Suzuki === **Michelia compressa** (Maxim.) Sarg.

Michelia foveolata var. *cinerascens* Y.W.Law & Y.F.Wu === **Michelia foveolata** Merr. ex Dandy

Michelia fulgens Dandy === **Michelia foveolata** Merr. ex Dandy

Michelia fuscata (Andrews) Blume === **Michelia figo** (Lour.) Spreng. var. **figo**

Michelia glaberrima H.T.Chang === **Michelia chapensis** Dandy

Michelia glabra P.Parm. === ?

Michelia glauca Wight === **Michelia nilagirica** Zenker

Michelia gracilis Kostel. === **Magnolia kobus** DC.

Michelia gravis Dandy ex Gagnep. === ?

Michelia griffithii (Hook.f. & Thomson) Finet & Gagnep. === **Magnolia griffithii** Hook.f. & Thomson

Michelia gustavii King === **Magnolia gustavii** King

Michelia hedyosperma Y.W.Law === **Michelia hypolampra** Dandy

Michelia iteophylla C.Y.Wu ex Y.W.Law & Y.F.Wu === **Michelia compressa** (Maxim.) Sarg.

Michelia kachirachirai Kaneh. & Yamam. === **Magnolia kachirachirai** (Kaneh. & Yamam.) Dandy

Michelia kerrii Craib === **Michelia floribunda** Finet & Gagnep.

Michelia lactea Buch.-Ham. ex Wall. === **Michelia oblonga** Wall. ex Hook.f. & Thomson

Michelia laevifolia Y.W.Law & Y.F.Wu === **Michelia yunnanensis** Franch. ex Finet & Gagnep.

Michelia lanceolata E.H.Wilson === **Michelia velutina** DC.

Michelia lanuginosa Wall. === **Michelia velutina** DC.

Michelia × *longifolia* Blume === **Michelia** × **alba** DC.

Michelia × *longifolia* var. *racemosa* Blume === **Michelia** × **alba** DC.

Michelia longipetiolata C.Y.Wu ex Y.W.Law & Y.F.Wu === **Michelia leveilleana** Dandy

Michelia longistamina Y.W.Law === **Michelia martinii** (H.Lév.) H.Lév.

Michelia longistyla Y.W.Law & Y.F.Wu === **Michelia aenea** Dandy

Michelia macclurei var. *sublanea* Dandy === **Michelia macclurei** Dandy

Michelia macrophylla D.Don === **Magnolia pterocarpa** Roxb.

Michelia magnifica Hu === **Michelia lacei** W.W.Sm.

Michelia manipurensis Watt ex Brandis === **Michelia doltsopa** Buch.-Ham. ex DC.

Michelia mediocris var. *angustifolia* G.A.Fu === **Michelia mediocris** Dandy

Michelia microcarpa B.L.Chen & S.C.Yang === **Michelia chapensis** Dandy

Michelia mollis (Dandy) McLaughlin === **Elmerrillia tsiampacca** subsp. **mollis** (Dandy) Noot.

Michelia montana var. *subvelutina* Miq. === **Michelia montana** Blume

Michelia nilagirica var. *walkeri* (Wight) Hook.f. & Thomson === **Michelia nilagirica** Zenker

Michelia nilagirica var. *wightii* Hook.f. & Thomson === **Michelia nilagirica** Zenker

Michelia nitida B.L.Chen === **Michelia chapensis** Dandy

Michelia oblongifolia Hung T.Chang & B.L.Chen === **Michelia aenea** Dandy

Michelia opipara Hung T.Chang & B.L.Chen === **Michelia doltsopa** Buch.-Ham. ex DC.

Michelia ovalifolia Wight === **Michelia nilagirica** Zenker

Michelia pachycarpa Y.W.Law & R.Z.Zhou === **Michelia lacei** W.W.Sm.

Michelia parviflora Merr. === **Michelia compressa** (Maxim.) Sarg.

Michelia parviflora Deless. === **Michelia figo** (Lour.) Spreng. var. **figo**

Michelia parviflora Rumph. ex DC. === ?

Michelia parvifolia (DC.) B.D.Jacks. === **Michelia figo** (Lour.) Spreng. var. **figo**

Michelia parvifolia Blume === ?

Michelia pealiana (King) Finet & Gagnep. === **Magnolia pealiana** King

Michelia pealiana King === **Magnolia pealiana** King

Michelia phellocarpa (King) Finet & Gagnep. === **Michelia baillonii** (Pierre) Finet & Gagnep.

Michelia philippinensis (P.Parm.) Dandy === **Michelia compressa** (Maxim.) Sarg.

Michelia pilifera Bakh.f. === **Michelia champaca** var. **pubinervia** (Blume) Miq.

Michelia platypetala Hand.-Mazz. === **Michelia cavaleriei** Finet & Gagnep.

Michelia platyphylla Merr. === **Elmerrillia platyphylla** (Merr.) Noot.

Michelia polyneura C.Y.Wu ex Y.W.Law & Y.F.Wu === **Michelia coriacea** Hung T.Chang & B.L.Chen

Michelia pubinervia Blume === **Michelia champaca** var. **pubinervia** (Blume) Miq.

Michelia pulneyensis Wight === **Michelia nilagirica** Zenker

Michelia rheedei Wight === **Michelia champaca** L. var. **champaca**

Michelia rufinervis DC. === **Michelia champaca** L. var. **champaca**

Michelia sericea Pers. === **Michelia champaca** L. var. **champaca**

Michelia sinensis Hemsl. & E.H.Wilson === **Michelia wilsonii** Finet & Gagnep.

Michelia skinneriana Dunn === **Michelia figo** (Lour.) Spreng. var. **figo**

Michelia sphaerantha C.Y.Wu ex Z.S.Yue === ?

Michelia sphaerantha C.Y.Wu ex Y.W.Law & Y.F.Wu === **Michelia masticata** Dandy

Michelia suaveolens Pers. === **Michelia champaca** L. var. **champaca**

Michelia sumatrae Dandy === **Michelia salicifolia** A.Agostini

Michelia szechuanica Dandy === **Michelia wilsonii** Finet & Gagnep.

Michelia tignifera Dandy === **Michelia lacei** W.W.Sm.

Michelia tila Buch.-Ham. ex Wall. === ?

Michelia tonkinensis A.Chev. === **Michelia balansae** (A.DC.) Dandy

Michelia tsiampacca L. === **Elmerrillia tsiampacca** (L.) Dandy

Michelia tsiampacca var. *blumei* Moritzi === **Michelia champaca** L. var. **champaca**

Michelia tsoi Dandy === **Michelia chapensis** Dandy
Michelia uniflora Dandy === **Michelia lacei** W.W.Sm.
Michelia velutina Blume === **Michelia champaca** var. **pubinervia** (Blume) Miq.
Michelia walkeri Wight === **Michelia nilagirica** Zenker
Michelia wardii Dandy === **Michelia doltsopa** Buch.-Ham. ex DC.
Michelia yulan (Desf.) Kostel. === **Magnolia denudata** Desr.
Michelia yunnanensis var. *angustifolia* Finet & Gagnep. === **Michelia yunnanensis** Franch. ex
 Finet & Gagnep.
Michelia zila Buch.-Ham. ex Madden === **Michelia kisopa** Buch.-Ham. ex DC.

Micheliopsis

Synonyms:
Micheliopsis H.Keng === **Magnolia** L.
Micheliopsis kachirachirai (Kaneh. & Yamam.) H.Keng === **Magnolia kachirachirai** (Kaneh.
 & Yamam.) Dandy

Pachylarnax

2 species, NE India, SE Asia and Malesia (Sumatra, Peninsular Malaysia). (Magnolieae)

- Nooteboom, H. P. (1985). Notes on Magnoliaceae. Blumea 31(1): 65-121. En. —
 Pachylarnax, pp. 97-98; revision (2 species), without key. *P. praecalva* is further treated in
 Flora Malesiana (Nooteboom 1988; see **Malesia**).

Pachylarnax Dandy, Bull. Misc. Inform. Kew 1927: 260 (1927).
 Trop. Asia. 40 41 42.

Pachylarnax pleiocarpa Dandy, J. Bot. 71: 313 (1933).
 Assam (Lakhimpur). 40 ASS. Phan.

Pachylarnax praecalva Dandy, Bull. Misc. Inform. Kew 1927: 260 (1927).
 Vietnam, Pen. Malaysia, W. Sumatra. 41 VIE 42 MLY SUM. Phan.

Parakmeria

Synonyms:
Parakmeria Hu & W.C.Cheng === **Magnolia** L.
Parakmeria kachirachirai (Kaneh. & Yamam.) Y.W.Law === **Magnolia kachirachirai** (Kaneh.
 & Yamam.) Dandy
Parakmeria kachirachirai (Kaneh. & Yamam.) Y.W.Law === **Magnolia kachirachirai** (Kaneh.
 & Yamam.) Dandy
Parakmeria nitida (W.W.Sm.) Y.W.Law === **Magnolia nitida** W.W.Sm. var. **nitida**
Parakmeria nitida (W.W.Sm.) Y.W.Law === **Magnolia nitida** W.W.Sm. var. **nitida**
Parakmeria omeiensis W.C.Cheng === **Magnolia omeiensis** (Hu & C.Y.Cheng) Dandy
Parakmeria yunnanensis Hu === **Magnolia nitida** W.W.Sm. var. **nitida**

Paramanglietia

Synonyms:
Paramanglietia Hu & W.C.Cheng === **Manglietia** Blume
Paramanglietia aromatica (Dandy) Hu & W.C.Cheng === **Manglietia aromatica** Dandy
Paramanglietia microcarpa H.T.Chang === **Manglietia fordiana** Oliv. var. **fordiana**

Paramichelia

Synonyms:
Paramichelia Hu === **Michelia** L.
Paramichelia baillonii (Pierre) Hu === **Michelia baillonii** (Pierre) Finet & Gagnep.
Paramichelia braianensis (Gagnep.) Dandy === **Michelia braianensis** Gagnep.
Paramichelia scortechinii (King) Dandy === **Michelia scortechinii** (King) Dandy

Sampacca

Synonyms:
Sampacca Kuntze === **Michelia** L.
Sampacca cathcartii (Hook.f. & Thomson) Kuntze === **Magnolia cathcartii** (Hook.f. & Thomson) Noot.
Sampacca euonymoides Kuntze === **Michelia champaca** L. var. **champaca**
Sampacca excelsa (Wall.) Kuntze === **Michelia doltsopa** Buch.-Ham. ex DC.
Sampacca kisopa (Buch.-Ham. ex DC.) Kuntze === **Michelia kisopa** Buch.-Ham. ex DC.
Sampacca lanuginosa (Wall.) Kuntze === **Michelia velutina** DC.
Sampacca × *longifolia* (Blume) Kuntze === **Michelia** × **alba** DC.
Sampacca montana (Blume) Kuntze === **Michelia montana** Blume
Sampacca nilagirica (Zenker) Kuntze === **Michelia nilagirica** Zenker
Sampacca oblonga (Wall. ex Hook.f. & Thomson) Kuntze === **Michelia oblonga** Wall. ex Hook.f. & Thomson
Sampacca parviflora (Deless.) Kuntze === **Michelia figo** (Lour.) Spreng. var. **figo**
Sampacca punduana (Hook.f. & Thomson) Kuntze === **Michelia punduana** Hook.f. & Thomson
Sampacca suaveolens (Pers.) Kuntze === **Michelia champaca** L. var. **champaca**
Sampacca velutina Kuntze === **Michelia champaca** L. var. **champaca**

Santanderia

Synonyms:
Santanderia Céspedes ex Triana & Planch. === **Magnolia** L.

Sinomanglietia

Synonyms:
Sinomanglietia Z.X.Yu === **Magnolia ?**
Sinomanglietia glauca Z.X.Yu === **Magnolia sp. ?**

Sphenocarpus

Synonyms:
Sphenocarpus Wall. === **Magnolia** L.
Sphenocarpus pterocarpus (Roxb.) K.Koch === **Magnolia pterocarpa** Roxb.

Svenhedinia

Synonyms:
Svenhedinia Urb. === **Magnolia** L.
Svenhedinia minor (Urb.) Urb. === **Magnolia minor** (Urb.) Govaerts
Svenhedinia truncata Moldenke === **Magnolia minor** (Urb.) Govaerts

Talauma

Synonyms:

Talauma A.Juss. === **Magnolia** L.

Talauma allenii (Standley) Lozano === **Magnolia allenii** Standl.

Talauma amazonica Ducke === **Magnolia amazonica** (Ducke) Govaerts

Talauma andamanica King === **Magnolia liliifera** (L.) Baill. var. **liliifera**

Talauma angatensis (Blanco) Fern.-Vill. === **Magnolia liliifera** var. **angatensis** (Blanco) Govaerts

Talauma arcabucoana Lozano === **Magnolia arcabucoana** (Lozano) Govaerts

Talauma athliantha Dandy === **Magnolia liliifera** (L.) Baill. var. **liliifera**

Talauma beccarii Ridl. === **Magnolia liliifera** var. **beccarii** (Ridl.) Govaerts

Talauma betongensis Craib === **Magnolia liliifera** var. **obovata** (Korth.) Govaerts

Talauma bintuluensis A.Agostini === **Magnolia bintuluensis** (A.Agostini) Noot.

Talauma boliviana M.Nee === **Magnolia boliviana** (M.Nee) Govaerts

Talauma borneensis Merr. === **Magnolia liliifera** (L.) Baill. var. **liliifera**

Talauma caerulea J.St.-Hil. === **Magnolia dodecapetala** (Lam.) Govaerts

Talauma candollei Blume === **Magnolia liliifera** (L.) Baill. var. **liliifera**

Talauma candollei var. *latifolia* Blume === **Magnolia liliifera** (L.) Baill. var. **liliifera**

Talauma caricifragrans Lozano === **Magnolia caricifragrans** (Lozano) Govaerts

Talauma celebica Koord. === ?

Talauma cespedesii Triana & Planch. === **Magnolia cespedesii** (Triana & Planch.) Govaerts

Talauma chocoensis Lozano === **Magnolia chocoensis** (Lozano) Govaerts

Talauma coco (Lour.) Merr. === **Magnolia coco** (Lour.) DC.

Talauma coerulea Steud. === **Magnolia dodecapetala** (Lam.) Govaerts

Talauma colombiana Little === **Magnolia colombiana** (Little) Govaerts

Talauma dixonii Little === **Magnolia dixonii** (Little) Govaerts

Talauma dodecapetala (Lam.) Urb. === **Magnolia dodecapetala** (Lam.) Govaerts

Talauma dubia Eichler === **Magnolia ovata** (A.St.-Hil.) Spreng.

Talauma duperreana (Pierre) Finet & Gagnep. === **Kmeria duperreana** (Pierre) Dandy

Talauma elegans (Blume) Miq. === **Magnolia elegans** (Blume) H.Keng

Talauma elegans var. *glauca* (Korth.) P.Parm. === **Magnolia elegans** (Blume) H.Keng

Talauma espinalii Lozano === **Magnolia espinalii** (Lozano) Govaerts

Talauma fistulosa Finet & Gagnep. === **Magnolia championii** Benth.

Talauma forbesii King === **Magnolia liliifera** (L.) Baill. var. **liliifera**

Talauma fragrantissima Hook. === **Magnolia sellowiana** (A.St.-Hil.) Govaerts

Talauma georgii Lozano === **Magnolia georgii** (Lozano) Govaerts

Talauma gigantifolia Miq. === **Magnolia gigantifolia** (Miq.) Noot.

Talauma gilbertoi Lozano === **Magnolia gilbertoi** (Lozano) Govaerts

Talauma gioii A.Chev. === **Michelia hypolampra** Dandy

Talauma gitingensis Elmer === **Magnolia liliifera** (L.) Baill. var. **liliifera**

Talauma gitingensis var. *glabra* Dandy === **Magnolia liliifera** (L.) Baill. var. **liliifera**

Talauma gitingensis var. *rotundata* Dandy === **Magnolia liliifera** (L.) Baill. var. **liliifera**

Talauma glaucum (Korth.) Miq. === **Magnolia elegans** (Blume) H.Keng

Talauma gloriensis Pittier === **Magnolia gloriensis** (Pittier) Govaerts

Talauma gracilior Dandy === **Magnolia liliifera** (L.) Baill. var. **liliifera**

Talauma grandiflora Merr. === **Magnolia liliifera** var. **angatensis** (Blanco) Govaerts

Talauma henaoi Lozano === **Magnolia henaoi** (Lozano) Govaerts

Talauma hernandezii Lozano === **Magnolia hernandezii** (Lozano) Govaerts

Talauma hodgsonii Hook.f. & Thomson === **Magnolia liliifera** var. **obovata** (Korth.) Govaerts

Talauma inflata P.Parm. === **Magnolia liliifera** (L.) Baill. var. **liliifera**

Talauma intonsa Dandy === **Magnolia sarawakensis** (A.Agostini) Noot.

Talauma irwiniana Lozano === **Magnolia irwiniana** (Lozano) Govaerts

Talauma javanica P.Parm. === **Magnolia liliifera** (L.) Baill. var. **liliifera**

Talauma katiorum Lozano === **Magnolia katiorum** (Lozano) Govaerts

Talauma kerrii Craib === **Magnolia henryi** Dunn

Talauma kunstleri King === **Magnolia liliifera** (L.) Baill. var. **liliifera**

Talauma kuteinensis A.Agostini === **Magnolia liliifera** var. **singapurensis** (Ridl.) Govaerts

Talauma lanigera Hook.f. & Thomson === **Magnolia villosa** (Miq.) H.Keng

Talauma levissima Dandy === **Magnolia liliifera** var. **obovata** (Korth.) Govaerts

Talauma liliifera (L.) Kuntze === **Magnolia liliifera** (L.) Baill.

Talauma liliifera Kurz === ?

Talauma longifolia (Blume) Ridl. === **Magnolia liliifera** (L.) Baill. var. **liliifera**

Talauma luzonensis Warb. === **Magnolia liliifera** var. **angatensis** (Blanco) Govaerts

Talauma macrocarpa Zucc. === **Magnolia mexicana** DC.

Talauma macrophylla Blume ex Miq. === **Magnolia liliifera** (L.) Baill. var. **liliifera**

Talauma magna A.Agostini === **Magnolia gigantifolia** (Miq.) Noot.

Talauma megalophylla Merr. === **Magnolia gigantifolia** (Miq.) Noot.

Talauma mexicana (DC.) G.Don === **Magnolia mexicana** DC.

Talauma minor Urb. === **Magnolia minor** (Urb.) Govaerts

Talauma minor var. *oblongifolia* Léon === **Magnolia minor** (Urb.) Govaerts

Talauma minor subsp. *oblongifolia* (Léon) Borhidi === **Magnolia minor** (Urb.) Govaerts

Talauma minor subsp. *orbiculata* (Britton & Wilson) Borhidi === **Magnolia minor** (Urb.) Govaerts

Talauma miqueliana Dandy === **Magnolia liliifera** (L.) Baill. var. **liliifera**

Talauma morii Lozano === **Magnolia morii** (Lozano) Govaerts

Talauma mutabilis Blume === **Magnolia liliifera** (L.) Baill. var. **liliifera**

Talauma mutabilis Fern.-Vill. === **Magnolia liliifera** var. **angatensis** (Blanco) Govaerts

Talauma mutabilis var. *acuminata* Blume === **Magnolia liliifera** (L.) Baill. var. **liliifera**

Talauma mutabilis var. *longifolia* Blume === **Magnolia liliifera** (L.) Baill. var. **liliifera**

Talauma mutabilis var. *splendens* (Urb.) Urb. ex McLaughlin === **Magnolia splendens** Urb.

Talauma mutabilis var. *splendens* Blume === **Magnolia liliifera** (L.) Baill. var. **liliifera**

Talauma narinensis Lozano === **Magnolia narinensis** (Lozano) Govaerts

Talauma neillii Lozano === **Magnolia neillii** (Lozano) Govaerts

Talauma nhatrangensis Dandy === **Magnolia liliifera** (L.) Baill. var. **liliifera**

Talauma oblanceolata Ridl. === **Magnolia liliifera** var. **obovata** (Korth.) Govaerts

Talauma oblongata Merr. === **Magnolia liliifera** var. **angatensis** (Blanco) Govaerts

Talauma oblongifolia (Léon) Bisse === **Magnolia minor** (Urb.) Govaerts

Talauma obovata Korth. === **Magnolia liliifera** var. **obovata** (Korth.) Govaerts

Talauma obovata (Siebold & Zucc.) Benth. & Hook.f. ex Hance === **Magnolia kobus** DC.

Talauma ophiticola Bisse === **Magnolia minor** (Urb.) Govaerts

Talauma orbiculata Britton & Wilson === **Magnolia minor** (Urb.) Govaerts

Talauma oreadum Diels === **Magnolia liliifera** (L.) Baill. var. **liliifera**

Talauma ovalis Miq. === **Elmerrillia ovalis** (Miq.) Dandy

Talauma ovata A.St.-Hil. === **Magnolia ovata** (A.St.-Hil.) Spreng.

Talauma papuana Schltr. === **Elmerrillia tsiampacca** (L.) Dandy subsp. **tsiampacca**

Talauma peninsularis Dandy === **Magnolia liliifera** (L.) Baill. var. **liliifera**

Talauma persuaveolens (Dandy) Dandy === **Magnolia persuaveolens** Dandy

Talauma phellocarpa King === **Michelia baillonii** (Pierre) Finet & Gagnep.

Talauma plumieri (Sw.) DC. === **Magnolia dodecapetala** (Lam.) Govaerts

Talauma poasana Pittier === **Magnolia poasana** (Pittier) Dandy

Talauma polyhypsophylla Lozano === **Magnolia polyhypsophylla** (Lozano) Govaerts

Talauma pubescens Merr. === **Elmerrillia pubescens** (Merr.) Dandy

Talauma pulgarensis Elmer === ?

Talauma pumila (Andrews) Blume === **Magnolia liliifera** (L.) Baill. var. **liliifera**

Talauma rabaniana Hook.f. & Thomson === **Magnolia liliifera** (L.) Baill. var. **liliifera**

Talauma rabaniana var. *villosa* (Miq.) P.Parm. === **Magnolia villosa** (Miq.) H.Keng

Talauma reticulata Merr. === **Magnolia liliifera** (L.) Baill. var. **liliifera**

Talauma rimachii Lozano === **Magnolia rimachii** (Lozano) Govaerts

Talauma roxburghii G.Don === **Magnolia pterocarpa** Roxb.

Talauma rubra Miq. === **Magnolia liliifera** (L.) Baill. var. **liliifera**

Talauma rumphii Blume === **Magnolia liliifera** (L.) Baill. var. **liliifera**

Talauma salicifolia Miq. === ?

Talauma salicifolia var. *concolor* Miq. === **Magnolia salicifolia** (Siebold & Zucc.) Maxim.

Talauma sambuensis Pittier === **Magnolia sambuensis** (Pittier) Govaerts

Talauma santanderiana Lozano === **Magnolia santanderiana** (Lozano) Govaerts

Talauma sarawakensis A.Agostini === **Magnolia sarawakensis** (A.Agostini) Noot.

Talauma sclerophylla Dandy === **Magnolia liliifera** var. **obovata** (Korth.) Govaerts

Talauma sebassa (King) Miq. ex Dandy === **Magnolia liliifera** (L.) Baill. var. **liliifera**

Talauma selloi (Spreng.) Steud. === **Magnolia sellowiana** (A.St.-Hil.) Govaerts

Talauma sellowiana A.St.-Hil. === **Magnolia sellowiana** (A.St.-Hil.) Govaerts

Talauma siamensis Dandy === **Magnolia liliifera** (L.) Baill. var. **liliifera**

Talauma sieboldii Miq. === **Magnolia liliiflora** Desr.

Talauma silvioi Lozano === **Magnolia silvioi** (Lozano) Govaerts

Talauma singapurensis Ridl. === **Magnolia liliifera** var. **singapurensis** (Ridl.) Govaerts

Talauma soembensis Dandy === **Magnolia liliifera** (L.) Baill. var. **liliifera**

Talauma splendens (Urb.) McLaughlin === **Magnolia splendens** Urb.

Talauma spongocarpa King === **Michelia baillonii** (Pierre) Finet & Gagnep.

Talauma stellata (Siebold & Zucc.) Miq. === **Magnolia stellata** (Siebold & Zucc.) Maxim.

Talauma sumatrana A.Agostini === **Magnolia liliifera** (L.) Baill. var. **liliifera**

Talauma thamnodes (Dandy) Tiep === **Magnolia liliifera** (L.) Baill. var. **liliifera**

Talauma truncata (Moldenke) R.A.Howard === **Magnolia minor** (Urb.) Govaerts

Talauma undulatifolia A.Agostini === **Magnolia liliifera** (L.) Baill. var. **liliifera**

Talauma venezuelensis Lozano === **Magnolia venezuelensis** (Lozano) Govaerts

Talauma villariana Rolfe === **Magnolia liliifera** var. **angatensis** (Blanco) Govaerts

Talauma villosa Miq. === **Magnolia villosa** (Miq.) H.Keng

Talauma villosa f. *celebica* Miq. === **Michelia champaca** var. **pubinervia** (Blume) Miq.

Talauma villosa f. *celebica* Miq. === **Michelia champaca** L. var. **champaca**

Talauma virolinensis Lozano === **Magnolia virolinensis** (Lozano) Govaerts

Talauma vrieseana Miq. === **Elmerrillia ovalis** (Miq.) Dandy

Talauma wolfii Lozano === **Magnolia wolfii** (Lozano) Govaerts

Tsoongiodendron

Synonyms:

Tsoongiodendron Chun === **Michelia** L.

Tsoongiodendron odorum Chun === **Michelia odora** (Chun) Noot. & B.L.Chen

Tulipastrum

Synonyms:

Tulipastrum Spach === **Magnolia** L.

Tulipastrum acuminatum (L.) Small === **Magnolia acuminata** (L.) L.

Tulipastrum acuminatum var. *aureum* Ashe === **Magnolia acuminata** (L.) L. var. **acuminata**

Tulipastrum acuminatum var. *ludovicianum* (Sarg.) Ashe === **Magnolia acuminata** (L.) L. var. **acuminata**

Tulipastrum americanum Spach === **Magnolia acuminata** (L.) L. var. **acuminata**

Tulipastrum americanum var. *subcordatum* Spach === **Magnolia acuminata** var. **subcordata** (Spach) Dandy

Tulipastrum americanum var. *vulgare* Spach === **Magnolia acuminata** (L.) L. var. **acuminata**

Tulipastrum cordatum (Michx.) Small === **Magnolia acuminata** var. **subcordata** (Spach) Dandy

Tulipifera

Synonyms:
Tulipifera Mill. === **Liriodendron** L.
Tulipifera liriodendrum Mill. === **Liriodendron tulipifera** L.

Yulania

Synonyms:
Yulania Spach === **Magnolia** L.
Yulania conspicua (Salisb.) Spach === **Magnolia denudata** Desr.
Yulania japonica Spach === **Magnolia liliiflora** Desr.
Yulania japonica var. *globosa* (Hook.f. & Thomson) P.Parm. === **Magnolia globosa** Hook.f. & Thomson
Yulania japonica var. *obovata* (Thunb.) P.Parm. === **Magnolia obovata** Thunb.
Yulania japonica var. *purpurea* (Curtis) P.Parm. === **Magnolia liliiflora** Desr.
Yulania kobus (DC.) Spach === **Magnolia kobus** DC.
Yulania × *lenneana* Lem. === **Magnolia** × **soulangeana** Soul.-Bod.

Appendix I

1. Summary of Insufficiently Known Taxa

Aromadendron yunnanense Hu, J. Roy. Hort. Soc. 63: 387 (1938).
=== ?

Champaca fasciculata Noronha, Verh. Batav. Genootsch. Kunsten, v. ed. I. Art.IV.: 13 (1790).
=== ?

Champaca turbinata Noronha, Verh. Batav. Genootsch. Kunsten, v. ed. I. Art.IV.: 12 (1790).
=== ?

Magnolia × **dorsopurpurea** Makino, J. Jap. Bot. 6(4): 8 (1929).
=== ?

Magnolia echinina P.Parm., Bull. Sc. France Belgique 27: 204, 265 (1896).
=== ?

Magnolia eriostepta Dandy ex Gagnep.P.H.Lecomte, Fl. Indo-Chine Suppl. 1: 39 (1938).
=== ?

Magnolia fasciculata P.Parm., Bull. Sc. France Belgique 27: 204, 265 (1896).
=== ?

Magnolia ferruginea P.Parm., Bull. Sc. France Belgique 27: 203, 263 (1896).
=== ?

Magnolia fragrans Raf., Fl. Ludov.: 91 (1817).
=== ?

Magnolia glabra P.Parm., Bull. Sc. France Belgique 27: 194, 251 (1896).
=== ?

Magnolia heliophyla P.Parm., Bull. Sc. France Belgique 27: 202, 262 (1896).
=== ?

Magnolia hirsuta Thunb., Pl. Jap. Nov. Sp.: 8 (1824).
=== ?

Magnolia × **hybrida** Dippel, Handb. Laubholzk. 3: 151 (1893).
=== ?

Magnolia inodora DC., Syst. Nat. 1: 459 (1817).
=== ?

Magnolia insignis Blume, Fl. Javae 19-20: 23 (1829).
=== ?

Magnolia intermedia P.Parm., Bull. Sc. France Belgique 27: 204, 266 (1896).
=== ?

Magnolia kobushii Mayr, Fremdländ. Wald- Parkbäume: 484 (1906).
=== ?

Magnolia longistyla P.Parm., Bull. Sc. France Belgique 27: 205, 267 (1896).
=== ?

Magnolia michauxii Fraser ex Thouin, Ann. Mus. Natl. Hist. Nat. 2: 252 (1803).
=== ?

Magnolia mutabilis Regel, Cat. Pl. Hort. Aksakov.: 88 (1860).
=== ?

Magnolia ovata P.Parm., Bull. Sc. France Belgique 27: 193, 250 (1896).
=== ?

Magnolia pulneyensis P.Parm., Bull. Sc. France Belgique 27: 205, 268 (1896).
=== ?

Magnolia sericea Thunb., Pl. Jap. Nov. Sp.: 8 (1824).
=== ?

Magnolia spathulata W.C.Cheng ex C.Pei, Proc. Fifth Pacific Sci. Congr. Canada 4: 3168 (1933 publ. 1934).
=== ?

Magnolia velutina P.Parm., Bull. Sc. France Belgique 27: 205, 269 (1896).
=== ?

Magnolia virginiana var. **grisea** L., Sp. Pl.: 536 (1753).
=== ?

Manglietia pachyphylla H.T.Chang, Acta Sci. Nat. Univ. Sunyatseni 1961: 55 (1961).
=== ?

Michelia caerulea DC., Syst. Nat. 1: 449 (1817).
=== ?

Michelia caloptila Y.W.Law & Y.F.Wu, Bull. Bot. Res., Harbin 4: 152 (1984).
=== ?

Michelia glabra P.Parm., Bull. Sc. France Belgique 27: 213, 282 (1896).
=== ?

Michelia gravis Dandy ex Gagnep.P.H.Lecomte, Fl. Indo-Chine Suppl. 1: 50 (1938).
=== ?

Michelia parviflora Rumph. ex DC., Syst. Nat. 1: 449 (1818).
=== ?

Michelia parvifolia Blume, Fl. Javae 19-20: 8 (1829).
=== ?

Michelia sphaerantha C.Y.Wu ex Z.S.Yue, Acta Bot. Yunnan. 9: 413 (1987).
=== ?

Michelia tila Buch.-Ham. ex Wall., Numer. List: 236 No. 970 B (1829).
=== ?

Sinomanglietia Z.X. Yu, Acta Agric. Univ. Jiangxiensis 16: 202 (1994).
=== **Magnolia** ?

Sinomanglietia glauca Z.X. Yu, Acta Agric. Univ. Jiangxiensis 16: 203 (1994).
=== **Magnolia sp.** ?

Talauma celebica Koord., Meded. Lands Plantentuin 19: 632 (1898).
=== ?

Talauma liliifera Kurz, J. Asiat. Soc. Bengal 43(2): 47 (1874).
=== ?

Talauma pulgarensis Elmer, Leafl. Philipp. 5: 1809 (1913).
=== ?

Talauma salicifolia Miq., Ann. Mus. Bot. Lugduno-Batavi 2: 258 (1866).
=== ?

2. Excluded taxa

Magnolia tomentosa Thunb., Trans. Linn. Soc. London 2: 336 (1794).
=== Edgeworthia papyrifera Sieb. & Zucc. (Thymelaeaceae)

Magnolia xerophila P. Parm., Bull. Sci. France Belgique 27: 203, 263 (1896).
=== Mimusops elengi L. (Sapotaceae)

Michelia acris Ruiz & Pav. ex Lopez, Anales Inst. Bot. Cavanilles 17: 415 (1959).
=== Drimys winteri J.R. Forst. & G. Forst. (Winteraceae)

Appendix II

Combinationium novarum summarium

The following is a summary list of all new combinations made in the text. These combinations are the result of discussions among specialists and with other interested persons at an international symposium, 'Magnolias and their allies', a joint undertaking of the International Dendrology Society and the Magnolia Society held at Royal Holloway, Egham, Surrey, United Kingdom, on 12-13 April 1996.

Magnolia amazonica (Ducke) Govaerts, comb. nov.
    Basionym: Talauma amazonica Ducke, Arch. Jard. Bot. Rio de Janeiro 4: 11 (1925).

Magnolia arcabucoana (Lozano-Contreras) Govaerts, comb. nov.
    Basionym: Talauma arcabucoana Lozano-Contreras, Fl. Colombia 1: 58 (1983).

Magnolia argyrotricha (Lozano-Contreras) Govaerts, comb. nov.
    Basionym: Dugandiodendron argyrotrichum Lozano-Contreras, Caldasia 11(53): 38 (1975).

Magnolia calimaensis (Lozano) Govaerts, comb. nov.
    Basionym: Dugandiodendron calimaense Lozano, Dugandiodendron Talauma Neotróp.: 35 (1994).

Magnolia calophylla (Lozano-Contreras) Govaerts, comb. nov.
    Basionym: Dugandiodendron calophyllum Lozano-Contreras, Caldasia 12(58): 283 (1978).

Magnolia cararensis (Lozano) Govaerts, comb. nov.
    Basionym: Dugandiodendron cararense Lozano, Dugandiodendron Talauma Neotróp.: 52 (1994).

Magnolia caricifragrans (Lozano-Contreras) Govaerts, comb. nov.
    Basionym: Talauma caricifragrans Lozano-Contreras, Mutisia 36: 2 (1972).

Magnolia cespedesii (Triana & Planch.) Govaerts, comb. nov.
    Basionym: Talauma cespedesii Triana & Planch., Ann. Sc. Nat. IV, 17: 23 (1862).

Magnolia chocoensis (Lozano-Contreras) Govaerts, comb. nov.
    Basionym: Talauma chocoensis Lozano-Contreras, Fl. Colombia 1: 67 (1983).

Magnolia colombiana (Little) Govaerts, comb. nov.
    Basionym: Talauma colombiana Little, Phytologia 19: 292 (1970).

Magnolia dixonii (Little) Govaerts, comb. nov.
    Basionym: Talauma dixonii Little, Phytologia 18: 457 (1969).

Magnolia dodecapetala (Lam.) Govaerts, comb. nov.
    Basionym: Annona dodecapetala Lam., Encycl. 2: 127 (1786).

Magnolia espinalii (Lozano-Contreras) Govaerts, comb. nov.
    Basionym: Talauma espinalii Lozano-Contreras, Fl. Colombia 1: 70 (1983).

Magnolia georgii (Lozano-Contreras) Govaerts, comb. nov.
    Basionym: Talauma georgii Lozano-Contreras, Fl. Colombia 1: 76 (1983).

Magnolia gilbertoi (Lozano-Contreras) Govaerts, comb. nov.
    Basionym: Talauma gilbertoi Lozano-Contreras, Fl. Colombia 1: 73 (1983).

Magnolia gloriensis (Pittier) Govaerts, comb. nov.
 Basionym: Talauma gloriensis Pittier, Contrib. U.S. Nat. Herb. 13: 94 (1910).

Magnolia guatapensis (Lozano) Govaerts, comb. nov.
 Basionym: Dugandiodendron guatapense Lozano, Dugandiodendron Talauma Neotróp.: 50 (1994).

Magnolia henaoi (Lozano-Contreras) Govaerts, comb. nov.
 Basionym: Talauma henaoi Lozano-Contreras, Fl. Colombia 1: 78 (1983).

Magnolia hernandezii (Lozano-Contreras) Govaerts, comb. nov.
 Basionym: Talauma hernandezii Lozano-Contreras, Mutisia 37: 11 (1972).

Magnolia irwiniana (Lozano-Contreras) Govaerts, comb. nov.
 Basionym: Talauma irwiniana Lozano-Contreras, Rev. Acad. Colomb. Cienc. Exact. Fis. Nat. 66: 580 (1990).

Magnolia katiorum (Lozano-Contreras) Govaerts, comb. nov.
 Basionym: Talauma katiorum Lozano-Contreras, Fl. Colombia 1: 84 (1983).

Magnolia lenticellatum (Lozano) Govaerts, comb. nov.
 Basionym: Dugandiodendron lenticellata Lozano, Dugandiodendron Talauma Neotróp.: 46 (1994).

Magnolia liliifera var. **angatensis** (Blanco) Govaerts, comb. nov.
 Basionym: Magnolia angatensis Blanco, Fl. Filip.: 859 (1837).

Magnolia liliifera var. **beccarii** (Ridley) Govaerts, comb. nov.
 Basionym: Talauma beccarii Ridley, Bull. Misc. Inform. Kew 1912: 381 (1912).

Magnolia liliifera var. **obovata** (Korth.) Govaerts, comb. nov.
 Basionym: Talauma obovata Korth., Ned. Kruidk. Arch. 2(2): 89 (1851).

Magnolia liliifera var. **singapurensis** (Ridley) Govaerts, comb. nov.
 Basionym: Talauma singapurensis Ridley, Bull. Misc. Inform. Kew 1914: 323 (1914).

Magnolia magnifolia (Lozano-Contreras) Govaerts, comb. nov.
 Basionym: Dugandiodendron magnifolium Lozano-Contreras, Fl. Colombia 1: 37 (1983).

Magnolia mahechae (Lozano-Contreras) Govaerts, comb. nov.
 Basionym: Dugandiodendron mahechae Lozano-Contreras, Caldasia 11(53): 33 (1975).

Magnolia minor (Urb.) Govaerts, comb. nov.
 Basionym: Talauma minor Urb., Symb. Antill. 7: 222 (1912).

Magnolia morii (Lozano) Govaerts, comb. nov.
 Basionym: Talauma morii Lozano, Dugandiodendron Talauma Neotróp.: 113 (1994).

Magnolia narinensis (Lozano-Contreras) Govaerts, comb. nov.
 Basionym: Talauma narinensis Lozano-Contreras, Caldasia 12(58): 286 (1978).

Magnolia neillii (Lozano) Govaerts, comb. nov.
 Basionym: Talauma neillii Lozano, Dugandiodendron Talauma Neotróp.: 71 (1994).

Magnolia polyhypsophylla (Lozano-Contreras) Govaerts, comb. nov.
 Basionym: Talauma polyhypsophylla Lozano-Contreras, Fl. Colombia 1: 87 (1983).

Magnolia rimachii (Lozano) Govaerts, comb. nov.
 Basionym: Talauma rimachii Lozano, Dugandiodendron Talauma Neotróp.: 105 (1994).

Magnolia sambuensis (Pittier) Govaerts, comb. nov.
Basionym: Talauma sambuensis Pittier, Contrib. U.S. Nat. Herb. 20: 105 (1918).

Magnolia santanderiana (Lozano-Contreras) Govaerts, comb. nov.
Basionym: Talauma santanderiana Lozano-Contreras, Fl. Colombia 1: 95 (1983).

Magnolia sellowiana (A. St. Hil.) Govaerts, comb. nov.
Basionym: Talauma sellowiana A. St. Hil., Fl. Bras. Mer. 1: 26 (1824).

Magnolia silvioi (Lozano-Contreras) Govaerts, comb. nov.
Basionym: Talauma silvioi Lozano-Contreras, Fl. Colombia 1: 98 (1983).

Magnolia urraoense (Lozano-Contreras) Govaerts, comb. nov.
Basionym: Dugandiodendron urraoense Lozano-Contreras, Fl. Colombia 1: 42 (1983).

Magnolia venezuelensis (Lozano) Govaerts, comb. nov.
Basionym: Talauma venezuelensis Lozano, Revista Acad. Colomb. Ci. Exact. 17: 78 (1990).

Magnolia virolinensis (Lozano-Contreras) Govaerts, comb. nov.
Basionym: Talauma virolinensis Lozano-Contreras, Fl. Colombia 1: 102 (1983).

Magnolia wolfii (Lozano) Govaerts, comb. nov.
Basionym: Talauma wolfii Lozano, Dugandiodendron Talauma Neotróp.: 90 (1994).

Magnolia yarumalense (Lozano-Contreras) Govaerts, comb. nov.
Basionym: Dugandiodendron yarumalense Lozano-Contreras, Fl. Colombia 1: 46 (1983).